A BRIEF HISTORY OF TIME

'Stephen Hawking affords a breathtaking insight, not just into the past, present and future of the Universe, but into the mind of scientific genius. Humble and often amusing, he puts profound ideas across in lucid terms laymen can understand'

Evening Standard

'He has certainly done more to make the idea of a big bang and the beginning of the universe coherent and thinkable than any physicist since Einstein . . . The reader can't help but imagine that if Einstein had tried to write a book for non-physicists, it might have been like this'

Newsday

'Mr Hawking clearly possesses a natural teacher's gifts – easy, good-natured humor and an ability to illustrate highly complex propositions with analogies plucked from daily life'

New York Times

'Our greatest theoretical physicist'

Independent

'The best single book on astrophysics for the common reader'
Booklist

'Stephen Hawking has overcome a crippling disease to become the supernova of world physics. Unable to write, or even to speak clearly, he is leaping beyond relativity, beyond quantum mechanics, beyond the big bang, to the "dance of geometry" that created the universe'
Vanity Fair

'*A Brief History of Time* lights up at such moments, when Mr Hawking allows us a peek at his impish humor, inner motivations, theoretical goofs, and scientific prejudices'
New York Times Book Review

'When the achievements of the physicists of the twentieth century come to be considered objectively, the work of Stephen Hawking will be writ large in the annals of science'
Astronomy

'This book is a *tour de force*'
The Scotsman

A BRIEF HISTORY OF TIME

FROM THE BIG BANG TO BLACK HOLES

STEPHEN W. HAWKING

INTRODUCTION BY CARL SAGAN
ILLUSTRATIONS BY RON MILLER

BANTAM BOOKS
TORONTO · NEW YORK · LONDON · SYDNEY · AUCKLAND

This book is dedicated to Jane

A BRIEF HISTORY OF TIME

A BANTAM BOOK 0 553 40165 3

Originally published in Great Britain by
Bantam Press, a division of Transworld Publishers Ltd.
PRINTING HISTORY
Bantam Press edition published 1988
Bantam edition published 1989
Reprinted November 1989
Reprinted January 1990
Reprinted March 1990
Reprinted August 1990
Reprinted October 1990
Bantam Books are published by Transworld Publishers Ltd.,
61–63 Uxbridge Road, Ealing, London W5 5SA, in
Australia by Transworld Publishers (Australia) Pty. Ltd.,
15–25 Helles Avenue, Moorebank, NSW 2170, and in New
Zealand by Transworld Publishers (N.Z.) Ltd., Cnr. Moselle
and Waipareira Avenues, Henderson, Auckland.

Printed and bound in Australia by The Book Printer, Victoria

Contents

Acknowledgments   vi
Introduction by Carl Sagan   ix
1 Our Picture of the Universe   1
2 Space and Time   15
3 The Expanding Universe   37
4 The Uncertainty Principle   57
5 Elementary Particles and the Forces of Nature   67
6 Black Holes   85
7 Black Holes Ain't So Black   105
8 The Origin and Fate of the Universe   121
9 The Arrow of Time   151
10 The Unification of Physics   163
11 Conclusion   181
Albert Einstein   187
Galileo Galilei   189
Isaac Newton   191
Glossary   193
Index   201

Acknowledgments

I decided to try and write a popular book about space and time after I gave the Loeb lectures at Harvard in 1982. There were already a considerable number of books about the early universe and black holes, ranging from the very good, such as Steven Weinberg's book, *The First Three Minutes,* to the very bad, which I will not identify. However, I felt that none of them really addressed the questions that had led me to do research in cosmology and quantum theory: Where did the universe come from? How and why did it begin? Will it come to an end, and if so, how? These are questions that are of interest to us all. But modern science has become so technical that only a very small number of specialists are able to master the mathematics used to describe them. Yet the basic ideas about the origin and fate of the universe can be stated without mathematics in a form that people without a scientific education can understand. This is what I have attempted to do in this book. The reader must judge whether I have succeeded.

Someone told me that each equation I included in the book

would halve the sales. I therefore resolved not to have any equations at all. In the end, however, I *did* put in one equation, Einstein's famous equation, $E = mc^2$. I hope that this will not scare off half of my potential readers.

Apart from being unlucky enough to get ALS, or motor neuron disease, I have been fortunate in almost every other respect. The help and support I received from my wife, Jane, and my children, Robert, Lucy, and Timmy, have made it possible for me to lead a fairly normal life and to have a successful career. I was again fortunate in that I chose theoretical physics, because that is all in the mind. So my disability has not been a serious handicap. My scientific colleagues have without exception been most helpful.

In the first "classical" phase of my career, my principal associates and collaborators were Roger Penrose, Robert Geroch, Brandon Carter, and George Ellis. I'm grateful to them for the help they gave me, and for the work we did together. This phase was summed up by the book, *The Large Scale Structure of Spacetime,* which Ellis and I wrote in 1973. I would not advise readers of this book to consult that work for further information: it is highly technical, and quite unreadable. I hope that since then I have learned how to write in a manner that is easier to understand.

In the second "quantum" phase of my work, since 1974, my principal collaborators have been Gary Gibbons, Don Page, and Jim Hartle. I owe a lot to them, and to my research students, who have given me a great deal of help, in both the physical and theoretical senses of the word. Having to keep up with my students has been a great stimulation, and has, I hope, prevented me from getting stuck in a rut.

I have had a lot of help with this book from Brian Whitt, one of my students. I caught pneumonia in 1985, after I had written the first draft. I had to have a tracheostomy operation which removed my ability to speak, and made it almost impossible for me to communicate. I thought I would be unable to finish it. However, Brian not only helped me revise it, he also got me using a communications program called

Living Center which was donated to me by Walt Woltosz, of Words Plus Inc., in Sunnyvale, California. With this I can both write books and papers, and speak to people using a speech synthesizer donated by Speech Plus, also of Sunnyvale, California. The synthesizer and a small personal computer were mounted on my wheelchair by David Mason. This system has made all the difference: In fact I can communicate better now than before I lost my voice.

I have had suggestions of how to improve the book from a large number of people who have seen preliminary versions. In particular, Peter Guzzardi, my editor at Bantam Books, sent me pages and pages of comments and queries about points he felt that I had not explained properly. I must admit that I was rather irritated when I received his great list of things to be changed, but he was quite right. I'm sure that it is a better book as a result of his keeping my nose to the grindstone.

I'm very grateful to my assistants, Colin Williams, David Thomas, and Raymond Laflamme; my secretaries Judy Fella, Ann Ralph, Cheryl Billington, and Sue Masey; and my team of nurses. None of this would have been possible without the support for my research and medical expenses that has been supplied by Gonville and Caius College, the Science and Engineering Research Council, and by the Leverhulme, McArthur, Nuffield, and Ralph Smith Foundations. I'm very grateful to them.

—Stephen Hawking
20th October 1987

Introduction

We go about our daily lives understanding almost nothing of the world. We give little thought to the machinery that generates the sunlight that makes life possible, to the gravity that glues us to an Earth that would otherwise send us spinning off into space, or to the atoms of which we are made and on whose stability we fundamentally depend. Except for children (who don't know enough not to ask the important questions), few of us spend much time wondering why nature is the way it is; where the cosmos came from, or whether it was always here; if time will one day flow backward and effects precede causes; or whether there are ultimate limits to what humans can know. There are even children, and I have met some of them, who want to know what a black hole looks like; what is the smallest piece of matter; why we remember the past and not the future; how it is, if there was chaos early, that there is, apparently, order today; and why there *is* a universe.

In our society it is still customary for parents and teachers to answer most of these questions with a shrug, or with an appeal to vaguely recalled religious precepts. Some are un-comfortable with issues like these, because they so vividly expose the limitations of human understanding.

But much of philosophy and science has been driven by such inquiries. An increasing number of adults are willing to ask questions of this sort, and occasionally they get some astonishing answers. Equidistant from the atoms and the stars, we are expanding our exploratory horizons to embrace both the very small and the very large.

In the spring of 1974, about two years before the Viking spacecraft landed on Mars, I was at a meeting in England sponsored by the Royal Society of London to explore the question of how to search for extraterrestrial life. During a coffee break I noticed that a much larger meeting was being held in an adjacent hall, which out of curiosity I entered. I soon realized that I was witnessing an ancient rite, the inves-titure of new fellows into the Royal Society, one of the most ancient scholarly organizations on the planet. In the front row a young man in a wheelchair was, very slowly, signing his name in a book that bore on its earliest pages the signature of Isaac Newton. When at last he finished, there was a stirring ovation. Stephen Hawking was a legend even then.

Hawking is now the Lucasian Professor of Mathematics at Cambridge University, a post once held by Newton and later by P.A.M. Dirac, two celebrated explorers of the very large and the very small. He is their worthy successor. This, Hawking's first book for the nonspecialist, holds rewards of many kinds for the lay audience. As interesting as the book's wide-ranging contents is the glimpse it provides into the workings of its author's mind. In this book are lucid revela-tions on the frontiers of physics, astronomy, cosmology, and courage.

This is also a book about God . . . or perhaps about the absence of God. The word God fills these pages. Hawking embarks on a quest to answer Einstein's famous question about whether God had any choice in creating the universe. Hawking is attempting, as he explicitly states, to understand the mind of God. And this makes all the more unexpected the conclusion of the effort, at least so far: a universe with no edge in space, no beginning or end in time, and nothing for a Creator to do.

—Carl Sagan
Cornell University
Ithaca, New York

1

Our Picture of the Universe

A well-known scientist (some say it was Bertrand Russell) once gave a public lecture on astronomy. He described how the earth orbits around the sun and how the sun, in turn, orbits around the center of a vast collection of stars called our galaxy. At the end of the lecture, a little old lady at the back of the room got up and said: "What you have told us is rubbish. The world is really a flat plate supported on the back of a giant tortoise." The scientist gave a superior smile before replying, "What is the tortoise standing on?" "You're very clever, young man, very clever," said the old lady. "But it's turtles all the way down!"

Most people would find the picture of our universe as an infinite tower of tortoises rather ridiculous, but why do we think we know better? What do we know about the universe, and how do we know it? Where did the universe come from, and where is it going? Did the universe have a beginning, and if so, what happened *before* then? What is the nature of time? Will it ever come to an end? Recent breakthroughs in physics, made possible in part by fantastic new technologies, suggest answers to some of these longstanding questions. Someday these answers may seem as obvious to us as the earth orbiting the sun—or perhaps as ridiculous as a tower of tortoises. Only time (whatever that may be) will tell.

As long ago as 340 B.C. the Greek philosopher Aristotle, in his book *On the Heavens,* was able to put forward two good arguments for believing that the earth was a round sphere rather than a flat plate. First, he realized that eclipses of the moon were caused by the earth coming between the sun and the moon. The earth's shadow on the moon was always round, which would be true only if the earth was spherical. If the earth had been a flat disk, the shadow would have been elongated and elliptical, unless the eclipse always occurred at a time when the sun was directly under the center of the disk. Second, the Greeks knew from their travels that the North Star appeared lower in the sky when viewed in the south than it did in more northerly regions. (Since the North Star lies over the North Pole, it appears to be directly above an observer at the North Pole, but to someone looking from the equator, it appears to lie just at the horizon.) From the difference in the apparent position of the North Star in Egypt and Greece, Aristotle even quoted an estimate that the distance around the earth was 400,000 stadia. It is not known exactly what length a stadium was, but it may have been about 200 yards, which would make Aristotle's estimate about twice the currently accepted figure. The Greeks even had a third argument that the earth must be round, for why else does one first see the sails of a ship coming over the horizon, and only later see the hull?

Aristotle thought that the earth was stationary and that the sun, the moon, the planets, and the stars moved in circular orbits about the earth. He believed this because he felt, for mystical reasons, that the earth was the center of the universe, and that circular motion was the most perfect. This idea was elaborated by Ptolemy in the second century A.D. into a complete cosmological model. The earth stood at the center, surrounded by eight spheres that carried the moon, the sun, the stars, and the five planets known at the time, Mercury, Venus, Mars, Jupiter, and Saturn (Fig. 1.1). The planets themselves moved on smaller circles attached to their respective spheres in order to account for their rather complicated observed paths in the sky. The outermost sphere carried the so-called fixed stars, which always stay in the same positions relative to each other but which rotate together across the sky. What lay beyond the last sphere was never made very

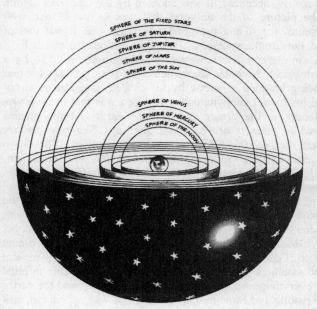

SPHERE OF THE FIXED STARS
SPHERE OF SATURN
SPHERE OF JUPITER
SPHERE OF MARS
SPHERE OF THE SUN
SPHERE OF VENUS
SPHERE OF MERCURY
SPHERE OF THE MOON

FIGURE 1.1

clear, but it certainly was not part of mankind's observable universe.

Ptolemy's model provided a reasonably accurate system for predicting the positions of heavenly bodies in the sky. But in order to predict these positions correctly, Ptolemy had to make an assumption that the moon followed a path that sometimes brought it twice as close to the earth as at other times. And that meant that the moon ought sometimes to appear twice as big as at other times! Ptolemy recognized this flaw, but nevertheless his model was generally, although not universally, accepted. It was adopted by the Christian church as the picture of the universe that was in accordance with Scripture, for it had the great advantage that it left lots of room outside the sphere of fixed stars for heaven and hell.

A simpler model, however, was proposed in 1514 by a Polish priest, Nicholas Copernicus. (At first, perhaps for fear of being branded a heretic by his church, Copernicus circulated his model anonymously.) His idea was that the sun was stationary at the center and that the earth and the planets moved in circular orbits around the sun. Nearly a century passed before this idea was taken seriously. Then two astronomers—the German, Johannes Kepler, and the Italian, Galileo Galilei—started publicly to support the Copernican theory, despite the fact that the orbits it predicted did not quite match the ones observed. The death blow to the Aristotelian/Ptolemaic theory came in 1609. In that year, Galileo started observing the night sky with a telescope, which had just been invented. When he looked at the planet Jupiter, Galileo found that it was accompanied by several small satellites or moons that orbited around it. This implied that everything did *not* have to orbit directly around the earth, as Aristotle and Ptolemy had thought. (It was, of course, still possible to believe that the earth was stationary at the center of the universe and that the moons of Jupiter moved on extremely complicated paths around the earth, giving the *appearance* that they orbited Jupiter. However, Copernicus's theory was much simpler.) At the same time, Johannes Kepler had modified Copernicus's theory, suggesting that the planets moved not in circles but in ellipses (an ellipse is an elongated circle). The predictions now finally matched the observations.

As far as Kepler was concerned, elliptical orbits were merely an ad hoc hypothesis, and a rather repugnant one at that, because ellipses were clearly less perfect than circles. Having discovered almost by accident that elliptical orbits fit the observations well, he could not reconcile them with his idea that the planets were made to orbit the sun by magnetic forces. An explanation was provided only much later, in 1687, when Sir Isaac Newton published his *Philosophiae Naturalis Principia Mathematica*, probably the most important single work ever published in the physical sciences. In it Newton not only put forward a theory of how bodies move in space and time, but he also developed the complicated mathematics needed to analyze those motions. In addition, Newton postulated a law of universal gravitation according to which each body in the universe was attracted toward every other body by a force that was stronger the more massive the bodies and the closer they were to each other. It was this same force that caused objects to fall to the ground. (The story that Newton was inspired by an apple hitting his head is almost certainly apocryphal. All Newton himself ever said was that the idea of gravity came to him as he sat "in a contemplative mood" and "was occasioned by the fall of an apple.") Newton went on to show that, according to his law, gravity causes the moon to move in an elliptical orbit around the earth and causes the earth and the planets to follow elliptical paths around the sun.

The Copernican model got rid of Ptolemy's celestial spheres, and with them, the idea that the universe had a natural boundary. Since "fixed stars" did not appear to change their positions apart from a rotation across the sky caused by the earth spinning on its axis, it became natural to suppose that the fixed stars were objects like our sun but very much farther away.

Newton realized that, according to his theory of gravity,

the stars should attract each other, so it seemed they could not remain essentially motionless. Would they not all fall together at some point? In a letter in 1691 to Richard Bentley, another leading thinker of his day, Newton argued that this would indeed happen if there were only a finite number of stars distributed over a finite region of space. But he reasoned that if, on the other hand, there were an infinite number of stars, distributed more or less uniformly over infinite space, this would not happen, because there would not be any central point for them to fall to.

This argument is an instance of the pitfalls that you can encounter in talking about infinity. In an infinite universe, every point can be regarded as the center, because every point has an infinite number of stars on each side of it. The correct approach, it was realized only much later, is to consider the finite situation, in which the stars all fall in on each other, and then to ask how things change if one adds more stars roughly uniformly distributed outside this region. According to Newton's law, the extra stars would make no difference at all to the original ones on average, so the stars would fall in just as fast. We can add as many stars as we like, but they will still always collapse in on themselves. We now know it is impossible to have an infinite static model of the universe in which gravity is always attractive.

It is an interesting reflection on the general climate of thought before the twentieth century that no one had suggested that the universe was expanding or contracting. It was generally accepted that either the universe had existed forever in an unchanging state, or that it had been created at a finite time in the past more or less as we observe it today. In part this may have been due to people's tendency to believe in eternal truths, as well as the comfort they found in the thought that even though they may grow old and die, the universe is eternal and unchanging.

Even those who realized that Newton's theory of gravity showed that the universe could not be static did not think to suggest that it might be expanding. Instead, they attempted to modify the theory by making the gravitational force repulsive at very large distances. This did not significantly affect their predictions of the motions of the planets, but it allowed an infinite distribution of stars to remain in equilibrium—with the attractive forces between nearby stars balanced by the repulsive forces from those that were farther away. However, we now believe such an equilibrium would be unstable: if the stars in some region got only slightly nearer each other, the attractive forces between them would become stronger and dominate over the repulsive forces so that the stars would continue to fall toward each other. On the other hand, if the stars got a bit farther away from each other, the repulsive forces would dominate and drive them farther apart.

Another objection to an infinite static universe is normally ascribed to the German philosopher Heinrich Olbers, who wrote about this theory in 1823. In fact, various contemporaries of Newton had raised the problem, and the Olbers article was not even the first to contain plausible arguments against it. It was, however, the first to be widely noted. The difficulty is that in an infinite static universe nearly every line of sight would end on the surface of a star. Thus one would expect that the whole sky would be as bright as the sun, even at night. Olbers's counterargument was that the light from distant stars would be dimmed by absorption by intervening matter. However, if that happened the intervening matter would eventually heat up until it glowed as brightly as the stars. The only way of avoiding the conclusion that the whole of the night sky should be as bright as the surface of the sun would be to assume that the stars had not been shining forever but had turned on at some finite time in the past. In that case the absorbing matter might not have heated up yet or the light from distant stars might not yet have reached us. And that brings us to the question of what could have caused the stars to have turned on in the first place.

The beginning of the universe had, of course, been discussed long before this. According to a number of early cosmologies and the Jewish/Christian/Muslim tradition, the universe started at a finite, and not very distant, time in the past. One argument for such a beginning was the feeling that it was necessary to have "First Cause" to explain the existence of the universe. (Within the universe, you always explained one event as being caused by some earlier event, but the existence of the universe itself could be explained in this way only if it had some beginning.) Another argument was put forward by St. Augustine in his book *The City of God*. He pointed out that civilization is progressing and we remember who performed this deed or developed that technique. Thus man, and so also perhaps the universe, could not have been around all that long. St. Augustine accepted a date of about 5000 B.C. for the Creation of the universe according to the Book of Genesis. (It is interesting that this is not so far from the end of the last Ice Age, about 10,000 B.C., which is when archaeologists tell us that civilization really began.)

Aristotle, and most of the other Greek philosophers, on the other hand, did not like the idea of a creation because it smacked too much of divine intervention. They believed, therefore, that the human race and the world around it had existed, and would exist, forever. The ancients had already considered the argument about progress described above, and answered it by saying that there had been periodic floods or other disasters that repeatedly set the human race right back to the beginning of civilization.

The questions of whether the universe had a beginning in time and whether it is limited in space were later extensively examined by the philosopher Immanuel Kant in his monumental (and very obscure) work, *Critique of Pure Reason*, published in 1781. He called these questions antinomies (that is, contradictions) of pure reason because he felt that there were equally compelling arguments for believing the thesis, that the universe had a beginning, and the antithesis, that it had existed forever. His argument for the thesis was that if the universe did not have a beginning, there would be an infinite period of time before any event, which he considered absurd. The argument for the antithesis was that if the universe had a beginning, there would be an infinite period of time before it, so why should the universe begin at any one particular time? In fact, his cases for both the thesis and the antithesis are really the same argument. They are both based on his unspoken assumption that time continues back forever, whether or not the universe had existed forever. As we shall see, the concept of time has no meaning before the beginning of the universe. This was first pointed out by St. Augustine. When asked: What did God do before he created the universe? Augustine didn't reply: He was preparing Hell for people who asked such questions. Instead, he said that time was a property of the universe that God created, and that time did not exist before the beginning of the universe.

When most people believed in an essentially static and unchanging universe, the question of whether or not it had a beginning was really one of metaphysics or theology. One could account for what was observed equally well on the theory that the universe had existed forever or on the theory that it was set in motion at some finite time in such a manner as to look as though it had existed forever. But in 1929, Edwin Hubble made the landmark observation that wherever you look, distant galaxies are moving rapidly away from us. In other words, the universe is expanding. This means that at earlier times objects would have been closer together. In fact, it seemed that there was a time, about ten or twenty thousand million years ago, when they were all at exactly the same place and when, therefore, the density of the universe was infinite. This discovery finally brought the question of the beginning of the universe into the realm of science.

Hubble's observations suggested that there was a time, called the big bang, when the universe was infinitesimally small and infinitely dense. Under such conditions all the laws of science, and therefore all ability to predict the future, would break down. If there were events earlier than this time, then they could not affect what happens at the present time. Their existence can be ignored because it would have no observational consequences. One may say that time had a beginning at the big bang, in the sense that earlier times simply would not be defined. It should be emphasized that this beginning in time is very different from those that had been considered previously. In an unchanging universe a beginning in time is something that has to be imposed by some being outside the universe; there is no physical necessity for a beginning. One can imagine that God created the universe at literally any time in the past. On the other hand, if the universe is expanding, there may be physical reasons why there had to be a beginning. One could still imagine that God created the universe at the instant of the big bang, or even afterwards in just such a way as to make it look as though there had been a big bang, but it would be meaningless to suppose that it was created *before* the big bang. An expanding universe does not preclude a creator, but it does place limits on when he might have carried out his job!

In order to talk about the nature of the universe and to discuss questions such as whether it has a beginning or an end, you have to be clear about what a scientific theory is. I shall take the simpleminded view that a theory is just a model of the universe, or a restricted part of it, and a set of rules that relate quantities in the model to observations that we make. It exists only in our minds and does not have any other reality (whatever that might mean). A theory is a good theory if it satisfies two requirements: It must accurately describe a large class of observations on the basis of a model that contains only a few arbitrary elements, and it must make definite predictions about the results of future observations. For example, Aristotle's theory that everything was made out of four elements, earth, air, fire, and water, was simple enough to qualify, but it did not make any definite predictions. On the other hand, Newton's theory of gravity was based on an even simpler model, in which bodies attracted each other with a force that was proportional to a quantity called their mass and inversely proportional to the square of

$F \propto \frac{m_1 + m_2}{r^2}$

the distance between them. Yet it predicts the motions of the sun, the moon, and the planets to a high degree of accuracy.

Any physical theory is always provisional, in the sense that it is only a hypothesis: you can never prove it. No matter how many times the results of experiments agree with some theory, you can never be sure that the next time the result will not contradict the theory. On the other hand, you can disprove a theory by finding even a single observation that disagrees with the predictions of the theory. As philosopher of science Karl Popper has emphasized, a good theory is characterized by the fact that it makes a number of predictions that could in principle be disproved or falsified by observation. Each time new experiments are observed to agree with the predictions the theory survives, and our confidence in it is increased; but if ever a new observation is found to disagree, we have to abandon or modify the theory. At least that is what is supposed to happen, but you can always question the competence of the person who carried out the observation.

In practice, what often happens is that a new theory is devised that is really an extension of the previous theory. For example, very accurate observations of the planet Mercury revealed a small difference between its motion and the predictions of Newton's theory of gravity. Einstein's general theory of relativity predicted a slightly different motion from Newton's theory. The fact that Einstein's predictions matched what was seen, while Newton's did not, was one of the crucial confirmations of the new theory. However, we still use Newton's theory for all practical purposes because the difference between its predictions and those of general relativity is very small in the situations that we normally deal with. (Newton's theory also has the great advantage that it is much simpler to work with than Einstein's!)

The eventual goal of science is to provide a single theory that describes the whole universe. However, the approach most scientists actually follow is to separate the problem into two parts. First, there are the laws that tell us how the universe changes with time. (If we know what the universe is like at any one time, these physical laws tell us how it will look at any later time.) Second, there is the question of the initial state of the universe. Some people feel that science should be concerned with only the first part; they regard the question of the initial situation as a matter for metaphysics or religion. They would say that God, being omnipotent, could have started the universe off any way he wanted. That may be so, but in that case he also could have made it develop in a completely arbitrary way. Yet it appears that he chose to make it evolve in a very regular way according to certain laws. It therefore seems equally reasonable to suppose that there are also laws governing the initial state.

It turns out to be very difficult to devise a theory to describe the universe all in one go. Instead, we break the problem up into bits and invent a number of partial theories. Each of these partial theories describes and predicts a certain limited class of observations, neglecting the effects of other quantities, or representing them by simple sets of numbers. It may be that this approach is completely wrong. If everything in the universe depends on everything else in a fundamental way, it might be impossible to get close to a full solution by investigating parts of the problem in isolation. Nevertheless, it is certainly the way that we have made progress in the past. The classic example again is the Newtonian theory of gravity, which tells us that the gravitational force between two bodies depends only on one number associated with each body, its mass, but is otherwise independent of what the bodies are made of. Thus one does not need to have a theory of the structure and constitution of the sun and the planets in order to calculate their orbits.

Today scientists describe the universe in terms of two basic partial theories—the general theory of relativity and quantum mechanics. They are the great intellectual achievements of the first half of this century. The general theory of relativity describes the force of gravity and the large-scale structure of the universe, that is, the structure on scales from only a few miles to as large as a million million million million (1 with twenty-four zeros after it) miles, the size of the observable universe. Quantum mechanics, on the other hand, deals with phenomena on extremely small scales, such as a millionth of a millionth of an inch. Unfortunately, however, these two theories are known to be inconsistent with each other—they cannot both be correct. One of the major endeavors in physics today, and the major theme of this book, is the search for a new theory that will incorporate them both—a quantum theory of gravity. We do not yet have such a theory, and we may still be a long way from having one, but we do already know many of the properties that it must have. And we shall see, in later chapters, that we already know a fair amount about the predictions a quantum theory of gravity must make.

Now, if you believe that the universe is not arbitrary, but is governed by definite laws, you ultimately have to combine the partial theories into a complete unified theory that will describe everything in the universe. But there is a fundamental paradox in the search for such a complete unified theory. The ideas about scientific theories outlined above assume we are rational beings who are free to observe the universe as we want and to draw logical deductions from what we see. In such a scheme it is reasonable to suppose that we might progress ever closer toward the laws that govern our universe. Yet if there really is a complete unified theory, it would also presumably determine our actions. And so the theory itself would determine the outcome of our search for it! And why should it determine that we come to the right conclusions from the evidence? Might it not equally well determine that we draw the wrong conclusion? Or no conclusion at all?

The only answer that I can give to this problem is based on Darwin's principle of natural selection. The idea is that in any population of self-reproducing organisms, there will be variations in the genetic material and upbringing that different individuals have. These differences will mean that some individuals are better able than others to draw the right conclusions about the world around them and to act accordingly.

These individuals will be more likely to survive and reproduce and so their pattern of behavior and thought will come to dominate. It has certainly been true in the past that what we call intelligence and scientific discovery has conveyed a survival advantage. It is not so clear that this is still the case: our scientific discoveries may well destroy us all, and even if they don't, a complete unified theory may not make much difference to our chances of survival. However, provided the universe has evolved in a regular way, we might expect that the reasoning abilities that natural selection has given us would be valid also in our search for a complete unified theory, and so would not lead us to the wrong conclusions.

Because the partial theories that we already have are sufficient to make accurate predictions in all but the most extreme situations, the search for the ultimate theory of the universe seems difficult to justify on practical grounds. (It is worth noting, though, that similar arguments could have been used against both relativity and quantum mechanics, and these theories have given us both nuclear energy and the microelectronics revolution!) The discovery of a complete unified theory, therefore, may not aid the survival of our species. It may not even affect our life-style. But ever since the dawn of civilization, people have not been content to see events as unconnected and inexplicable. They have craved an understanding of the underlying order in the world. Today we still yearn to know why we are here and where we came from. Humanity's deepest desire for knowledge is justification enough for our continuing quest. And our goal is nothing less than a complete description of the universe we live in.

2

Space and Time

Our present ideas about the motion of bodies date back to Galileo and Newton. Before them people believed Aristotle, who said that the natural state of a body was to be at rest and that it moved only if driven by a force or impulse. It followed that a heavy body should fall faster than a light one, because it would have a greater pull toward the earth.

The Aristotelian tradition also held that one could work out all the laws that govern the universe by pure thought: it was not necessary to check by observation. So no one until Galileo bothered to see whether bodies of different weight did in fact fall at different speeds. It is said that Galileo demonstrated that Aristotle's belief was false by dropping weights from the leaning tower of Pisa. The story is almost certainly untrue, but Galileo did do something equivalent: he rolled balls of different weights down a smooth slope. The situation is similar to that of heavy bodies falling vertically, but it is easier to observe because the speeds are smaller. Galileo's measurements indicated that each body increased its speed at the same rate, no matter what its weight. For example, if you let go of a ball on a slope that drops by one meter for every ten meters you go along, the ball will be traveling down the slope at a speed of about one meter per second after one second, two meters per second after two seconds, and so on, however heavy the ball. Of course a lead weight would fall faster than a feather, but that is only because a feather is slowed down by air resistance. If one drops two bodies that don't have much air resistance, such as two different lead weights, they fall at the same rate.

Galileo's measurements were used by Newton as the basis of his laws of motion. In Galileo's experiments, as a body rolled down the slope it was always acted on by the same force (its weight), and the effect was to make it constantly speed up. This showed that the real effect of a force is always to change the speed of a body, rather than just to set it moving, as was previously thought. It also meant that whenever a body is not acted on by any force, it will keep on moving in a straight line at the same speed. This idea was first stated explicitly in Newton's *Principia Mathematica*, published in 1687, and is known as Newton's first law. What happens to a body when a force does act on it is given by Newton's second law. This states that the body will accelerate, or change its speed, at a rate that is proportional to the force. (For example, the acceleration is twice as great if the force is twice as great.) The acceleration is also smaller the greater the mass (or quantity of matter) of the body. (The same force acting on a body of twice the mass will produce half the acceleration.) A familiar example is provided by a car: the more powerful the engine, the greater the acceleration, but the heavier the car, the smaller the acceleration for the same engine.

In addition to his laws of motion, Newton discovered a law to describe the force of gravity, which states that every body attracts every other body with a force that is proportional to the mass of each body. Thus the force between two bodies would be twice as strong if one of the bodies (say, body *A*) had its mass doubled. This is what you might expect because one could think of the new body *A* as being made of two bodies with the original mass. Each would attract body *B* with the original force. Thus the total force between *A* and *B* would be twice the original force. And if, say, one of the bodies had twice the mass, and the other had three times the mass, then the force would be six times as strong. One can now see why all bodies fall at the same rate: a body of twice the weight will have twice the force of gravity pulling it down, but it will also have twice the mass. According to Newton's second law, these two effects will exactly cancel each other, so the acceleration will be the same in all cases.

Newton's law of gravity also tells us that the farther apart the bodies, the smaller the force. Newton's law of gravity says that the gravitational attraction of a star is exactly one quarter that of a similar star at half the distance. This law predicts the orbits of the earth, the moon, and the planets with great accuracy. If the law were that the gravitational attraction of a star went down faster with distance, the orbits of the planets would not be elliptical, they would spiral in to the sun. If it went down slower, the gravitational forces from the distant stars would dominate over that from the earth.

The big difference between the ideas of Aristotle and those of Galileo and Newton is that Aristotle believed in a preferred state of rest, which any body would take up if it were not driven by some force or impulse. In particular, he thought that the earth was at rest. But it follows from Newton's laws that there is no unique standard of rest. One could equally well say that body *A* was at rest and body *B* was moving at constant speed with respect to body *A*, or that body *B* was at rest and body *A* was moving. For example, if one sets aside for a moment the rotation of the earth and its orbit round the sun, one could say that the earth was at rest and that a train on it was traveling north at ninety miles per hour or that the train was at rest and the earth was moving south at ninety miles per hour. If one carried out experiments with moving bodies on the train, all Newton's laws would still hold. For instance, playing Ping-Pong on the train, one would find that the ball obeyed Newton's laws just like a ball on a table by the track. So there is no way to tell whether it is the train or the earth that is moving.

The lack of an absolute standard of rest meant that one could not determine whether two events that took place at different times occurred in the same position in space. For example, suppose our Ping-Pong ball on the train bounces straight up and down, hitting the table twice on the same spot one second apart. To someone on the track, the two bounces would seem to take place about forty meters apart, because the train would have traveled that far down the track between the bounces. The nonexistence of absolute rest therefore meant that one could not give an event an absolute position in space, as Aristotle had believed. The positions of events and the distances between them would be different for a person on the train and one on the track, and there would be no reason to prefer one person's positions to the other's.

Newton was very worried by this lack of absolute position, or absolute space, as it was called, because it did not accord with his idea of an absolute God. In fact, he refused to accept lack of absolute space, even though it was implied by his laws. He was severely criticized for this irrational belief by many people, most notably by Bishop Berkeley, a philosopher who believed that all material objects and space and time are an illusion. When the famous Dr. Johnson was told of Berkeley's opinion, he cried, "I refute it thus!" and stubbed his toe on a large stone.

Both Aristotle and Newton believed in absolute time. That is, they believed that one could unambiguously measure the interval of time between two events, and that this time would be the same whoever measured it, provided they used a good clock. Time was completely separate from and independent of space. This is what most people would take to be the commonsense view. However, we have had to change our ideas about space and time. Although our apparently commonsense notions work well when dealing with things like apples, or planets that travel comparatively slowly, they don't work at all for things moving at or near the speed of light.

The fact that light travels at a finite, but very high, speed was first discovered in 1676 by the Danish astronomer Ole Christensen Roemer. He observed that the times at which the moons of Jupiter appeared to pass behind Jupiter were not evenly spaced, as one would expect if the moons went round Jupiter at a constant rate. As the earth and Jupiter orbit around the sun, the distance between them varies. Roemer noticed that eclipses of Jupiter's moons appeared later the farther we were from Jupiter. He argued that this was because the light from the moons took longer to reach us when we were farther away. His measurements of the variations in the distance of the earth from Jupiter were, however, not very accurate, and so his value for the speed of light was 140,000 miles per second, compared to the modern value of 186,000 miles per second. Nevertheless, Roemer's achievement, in not only proving that light travels at a finite speed, but also in measuring that speed, was remarkable—coming as it did eleven years before Newton's publication of *Principia Mathematica*.

A proper theory of the propagation of light didn't come until 1865 when the British physicist James Clerk Maxwell succeeded in unifying the partial theories that up to then had been used to describe the forces of electricity and magnetism. Maxwell's equations predicted that there could be wavelike disturbances in the combined electromagnetic field, and that these would travel at a fixed speed, like ripples on a pond. If the wavelength of these waves (the distance between one wave crest and the next) is a meter or more, they are what we now call radio waves. Shorter wavelengths are known as microwaves (a few centimeters) or infrared (more than a ten thousandth of a centimeter). Visible light has a wavelength of between only forty and eighty millionths of a centimeter. Even shorter wavelengths are known as ultraviolet, X rays, and gamma rays.

Maxwell's theory predicted that radio or light waves should

travel at a certain fixed speed. But Newton's theory had got rid of the idea of absolute rest, so if light was supposed to travel at a fixed speed, one would have to say what that fixed speed was to be measured relative to. It was therefore suggested that there was a substance called the "ether" that was present everywhere, even in "empty" space. Light waves should travel through the ether as sound waves travel through air, and their speed should therefore be relative to the ether. Different observers, moving relative to the ether, would see light coming toward them at different speeds, but light's speed relative to the ether would remain fixed. In particular, as the earth was moving through the ether on its orbit round the sun, the speed of light measured in the direction of the earth's motion through the ether (when we were moving toward the source of the light) should be higher than the speed of light at right angles to that motion (when we are not moving toward the source). In 1887 Albert Michelson (who later became the first American to receive the Nobel prize for physics) and Edward Morley carried out a very careful experiment at the Case School of Applied Science in Cleveland. They compared the speed of light in the direction of the earth's motion with that at right angles to the earth's motion. To their great surprise, they found they were exactly the same!

Between 1887 and 1905 there were several attempts, most notably by the Dutch physicist Hendrik Lorentz, to explain the result of the Michelson–Morley experiment in terms of objects contracting and clocks slowing down when they moved through the ether. However, in a famous paper in 1905, a hitherto unknown clerk in the Swiss patent office, Albert Einstein, pointed out that the whole idea of an ether was unnecessary, providing one was willing to abandon the idea of absolute time. A similar point was made a few weeks later by a leading French mathematician, Henri Poincaré. Einstein's arguments were closer to physics than those of Poincaré, who regarded this problem as mathematical. Einstein is usually given the credit for the new theory, but Poincaré is remembered by having his name attached to an important part of it.

The fundamental postulate of the theory of relativity, as it was called, was that the laws of science should be the same for all freely moving observers, no matter what their speed. This was true for Newton's laws of motion, but now the idea was extended to include Maxwell's theory and the speed of light: all observers should measure the same speed of light, no matter how fast they are moving. This simple idea has some remarkable consequences. Perhaps the best known are the equivalence of mass and energy, summed up in Einstein's famous equation $E = mc^2$ (where E is energy, m is mass and c is the speed of light), and the law that nothing may travel faster than the speed of light. Because of the equivalence of energy and mass, the energy which an object has due to its motion will add to its mass. In other words, it will make it harder to increase its speed. This effect is only really significant for objects moving at speeds close to the speed of light. For example, at 10 percent of the speed of light an object's mass is only 0.5 percent more than normal, while at 90 percent of the speed of light it would be more than twice its normal mass. As an object approaches the speed of light, its mass rises ever more quickly, so it takes more and more energy to speed it up further. It can in fact never reach the speed of light, because by then its mass would have become infinite, and by the equivalence of mass and energy, it would have taken an infinite amount of energy to get it there. For this reason, any normal object is forever confined by relativity to move at speeds slower than the speed of light. Only light, or other waves that have no intrinsic mass, can move at the speed of light.

An equally remarkable consequence of relativity is the way it has revolutionized our ideas of space and time. In Newton's theory, if a pulse of light is sent from one place to another, different observers would agree on the time that the journey took (since time is absolute), but will not always agree on how far the light traveled (since space is not absolute). Since the speed of the light is just the distance it has traveled divided by the time it has taken, different observers would measure different speeds for the light. In relativity, on the other hand, all observers *must* agree on how fast light travels. They still, however, do not agree on the distance the light has traveled, so they must therefore now also disagree over the time it has taken. (The time taken is the distance the light has traveled—which the observers do not agree on—divided by the light's speed—which they do agree on.) In other words, the theory of relativity put an end to the idea of absolute time! It appeared that each observer must have his own measure of time, as recorded by a clock carried with him, and that identical clocks carried by different observers would not necessarily agree.

Each observer could use radar to say where and when an event took place by sending out a pulse of light or radio waves. Part of the pulse is reflected back at the event and the observer measures the time at which he receives the echo. The time of the event is then said to be the time halfway between when the pulse was sent and the time when the reflection was received back: the distance of the event is half the time taken for this round trip, multiplied by the speed of light. (An event, in this sense, is something that takes place at a single point in space, at a specified point in time.) This idea is shown in Fig. 2.1, which is an example of a space-time diagram. Using this procedure, observers who are moving relative to each other will assign different times and positions to the same event. No particular observer's measurements are any more correct than any other observer's, but all the measurements are related. Any observer can work out precisely what time and position any other observer will assign to an event, provided he knows the other observer's relative velocity.

Nowadays we use just this method to measure distances precisely, because we can measure time more accurately than length. In effect, the meter is defined to be the distance traveled by light in 0.000000003335640952 seconds, as mea-

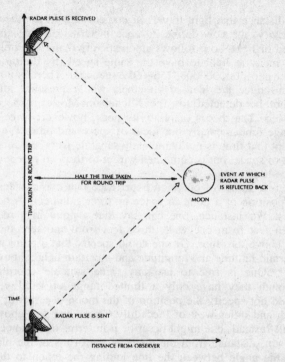

FIGURE 2.1 Time is measured vertically, and the distance from the observer is measured horizontally. The observer's path through space and time is shown as the vertical line on the left. The paths of light rays to and from the event are the diagonal lines.

sured by a cesium clock. (The reason for that particular number is that it corresponds to the historical definition of the meter—in terms of two marks on a particular platinum bar kept in Paris.) Equally, we can use a more convenient, new unit of length called a light-second. This is simply defined as the distance that light travels in one second. In the theory of relativity, we now define distance in terms of time and the speed of light, so it follows automatically that every observer will measure light to have the same speed (by definition, 1 meter per 0.000000003335640952 seconds). There is no need to introduce the idea of an ether, whose presence anyway cannot be detected, as the Michelson–Morley experiment showed. The theory of relativity does, however, force us to change fundamentally our ideas of space and time. We must accept that time is not completely separate from and independent of space, but is combined with it to form an object called space-time.

It is a matter of common experience that one can describe the position of a point in space by three numbers, or coordinates. For instance, one can say that a point in a room is seven feet from one wall, three feet from another, and five feet above the floor. Or one could specify that a point was at a certain latitude and longitude and a certain height above sea level. One is free to use any three suitable coordinates, although they have only a limited range of validity. One would not specify the position of the moon in terms of miles north and miles west of Piccadilly Circus and feet above sea level. Instead, one might describe it in terms of distance from the sun, distance from the plane of the orbits of the planets, and the angle between the line joining the moon to the sun and the line joining the sun to a nearby star such as Alpha Centauri. Even these coordinates would not be of much use in describing the position of the sun in our galaxy or the position of our galaxy in the local group of galaxies. In fact, one may describe the whole universe in terms of a collection of overlapping patches. In each patch, one can use a different set of three coordinates to specify the position of a point.

An event is something that happens at a particular point in space and at a particular time. So one can specify it by four numbers or coordinates. Again, the choice of coordinates is arbitrary; one can use any three well-defined spatial coordinates and any measure of time. In relativity, there is no real distinction between the space and time coordinates, just as there is no real difference between any two space coordinates. One could choose a new set of coordinates in which, say, the first space coordinate was a combination of the old first and second space coordinates. For instance, instead of measuring the position of a point on the earth in miles north of Piccadilly and miles west of Piccadilly, one could use miles northeast of Piccadilly, and miles northwest of Piccadilly. Similarly, in relativity, one could use a new time coordinate that was the old time (in seconds) plus the distance (in light-seconds) north of Piccadilly.

It is often helpful to think of the four coordinates of an event as specifying its position in a four-dimensional space called space-time. It is impossible to imagine a four-dimensional space. I personally find it hard enough to visualize three-dimensional space! However, it is easy to draw diagrams of two-dimensional spaces, such as the surface of the earth. (The surface of the earth is two-dimensional because the position of a point can be specified by two coordinates, latitude and longitude.) I shall generally use diagrams in which time increases upward and one of the spatial dimensions is shown horizontally. The other two spatial dimensions are ignored or, sometimes, one of them is indicated by perspective. (These are called space-time diagrams, like Fig. 2.1.) For example, in Fig. 2.2 time is measured upward in years and the distance along the line from the sun to Alpha Centauri is measured horizontally in miles. The paths of the sun and of Alpha Centauri through space-time are shown as the vertical lines on the left and right of the diagram. A ray of light from the sun follows the diagonal line, and takes four years to get from the sun to Alpha Centauri.

As we have seen, Maxwell's equations predicted that the speed of light should be the same whatever the speed of the source, and this has been confirmed by accurate measurements. It follows from this that if a pulse of light is emitted at a particular time at a particular point in space, then as time goes on it will spread out as a sphere of light whose size and

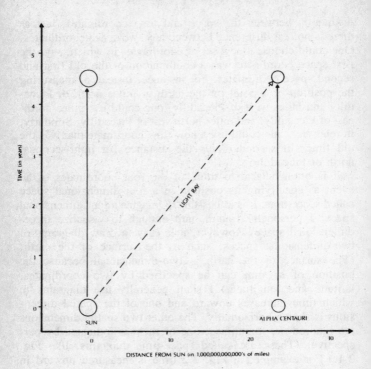

FIGURE 2.2

position are independent of the speed of the source. After one millionth of a second the light will have spread out to form a sphere with a radius of 300 meters; after two millionths of a second, the radius will be 600 meters; and so on. It will be like the ripples that spread out on the surface of a pond when a stone is thrown in. The ripples spread out as a circle that gets bigger as time goes on. If one thinks of a three-dimensional model consisting of the two-dimensional surface of the pond

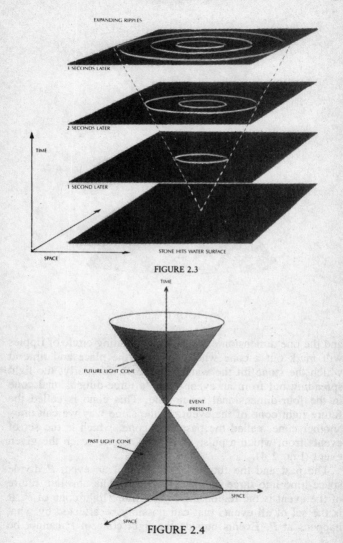

EXPANDING RIPPLES

4 SECONDS LATER

2 SECONDS LATER

1 SECOND LATER

TIME

SPACE

STONE HITS WATER SURFACE

FIGURE 2.3

TIME

FUTURE LIGHT CONE

EVENT (PRESENT)

PAST LIGHT CONE

SPACE

SPACE

FIGURE 2.4

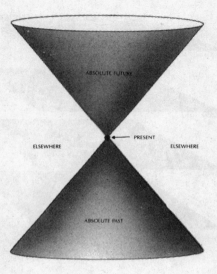

FIGURE 2.5

and the one dimension of time, the expanding circle of ripples will mark out a cone whose tip is at the place and time at which the stone hit the water (Fig. 2.3). Similarly, the light spreading out from an event forms a three-dimensional cone in the four-dimensional space-time. This cone is called the future light cone of the event. In the same way we can draw another cone, called the past light cone, which is the set of events from which a pulse of light is able to reach the given event (Fig. 2.4).

The past and the future light cones of an event P divide space-time into three regions (Fig. 2.5). The absolute future of the event is the region inside the future light cone of P. It is the set of all events that can possibly be affected by what happens at P. Events outside the light cone of P cannot be

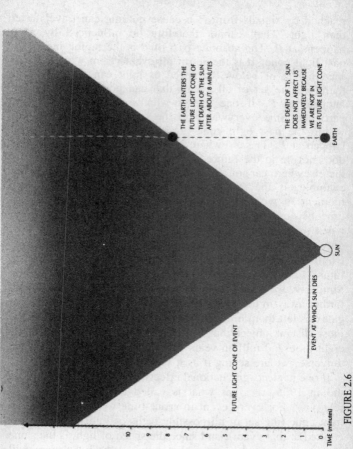

THE EARTH ENTERS THE
FUTURE LIGHT CONE OF
THE DEATH OF THE SUN
AFTER ABOUT 8 MINUTES

THE DEATH OF THE SUN
DOES NOT AFFECT US
IMMEDIATELY BECAUSE
WE ARE NOT IN
ITS FUTURE LIGHT CONE

EARTH

SUN

EVENT AT WHICH SUN DIES

FUTURE LIGHT CONE OF EVENT

TIME (minutes)

10
9
8
7
6
5
4
3
2
1
0

FIGURE 2.6

FIGURE 2.6

reached by signals from P because nothing can travel faster than light. They cannot therefore be influenced by what happens at P. The absolute past of P is the region inside the past light cone. It is the set of all events from which signals traveling at or below the speed of light can reach P. It is therefore the set of all events that can possibly affect what happens at P. If one knows what is happening at some particular time everywhere in the region of space that lies within the past light cone of P one can predict what will happen at P. The elsewhere is the region of space-time that does not lie in the future or past light cones of P. Events in the elsewhere cannot affect or be affected by events at P. For example, if the sun were to cease to shine at this very moment, it would not affect things on Earth at the present time because they would be in the elsewhere of the event when the sun went out (Fig. 2.6). We would know about it only after eight minutes, the time it takes light to reach us from the sun. Only then would events on Earth lie in the future light cone of the event at which the sun went out. Similarly, we do not know what is happening at the moment farther away in the universe: the light that we see from distant galaxies left them millions of years ago, and in the case of the most distant object that we have seen, the light left some eight thousand million years ago. Thus, when we look at the universe, we are seeing it as it was in the past.

If one neglects gravitational effects, as Einstein and Poincaré did in 1905, one has what is called the special theory of relativity. For every event in space-time we may construct a light cone (the set of all possible paths of light in space-time emitted at that event), and since the speed of light is the same at every event and in every direction, all the light cones will be identical and will all point in the same direction. The theory also tells us that nothing can travel faster than light. This means that the path of any object through space and time must be represented by a line that lies within the light cone at each event on it (Fig. 2.7).

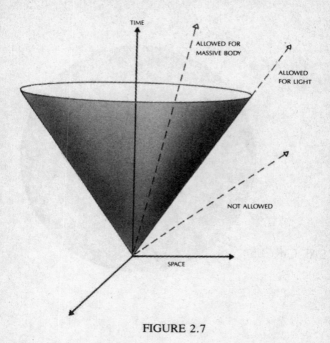

FIGURE 2.7

The special theory of relativity was very successful in explaining that the speed of light appears the same to all observers (as shown by the Michelson–Morley experiment) and in describing what happens when things move at speeds close to the speed of light. However, it was inconsistent with the Newtonian theory of gravity, which said that objects attracted each other with a force that depended on the distance between them. This meant that if one moved one of the objects, the force on the other one would change instantaneously. Or in other words, gravitational effects should travel with infinite velocity, instead of at or below the speed of light, as the special theory of relativity required. Einstein made a number of unsuccessful attempts between 1908 and

GREAT CIRCLE

FIGURE 2.8

1914 to find a theory of gravity that was consistent with special relativity. Finally, in 1915, he proposed what we now call the general theory of relativity.

Einstein made the revolutionary suggestion that gravity is not a force like other forces, but is a consequence of the fact that space-time is not flat, as had been previously assumed: it is curved, or "warped," by the distribution of mass and energy in it. Bodies like the earth are not made to move on curved orbits by a force called gravity; instead, they follow the nearest thing to a straight path in a curved space, which is called a geodesic. A geodesic is the shortest (or longest) path between two nearby points. For example, the surface of the earth is a two-dimensional curved space. A geodesic on the

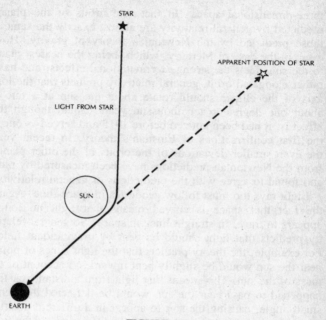

STAR

APPARENT POSITION OF STAR

LIGHT FROM STAR

SUN

EARTH

FIGURE 2.9

earth is called a great circle, and is the shortest route between two points (Fig. 2.8). As the geodesic is the shortest path between any two airports, this is the route an airline navigator will tell the pilot to fly along. In general relativity, bodies always follow straight lines in four-dimensional space-time, but they nevertheless appear to us to move along curved paths in our three-dimensional space. (This is rather like watching an airplane flying over hilly ground. Although it follows a straight line in three-dimensional space, its shadow follows a curved path on the two-dimensional ground.)

The mass of the sun curves space-time in such a way that although the earth follows a straight path in four-dimensional space-time, it appears to us to move along a circular orbit in three-dimensional space. In fact, the orbits of the planets predicted by general relativity are almost exactly the same as those predicted by the Newtonian theory of gravity. However, in the case of Mercury, which, being the nearest planet to the sun, feels the strongest gravitational effects, and has a rather elongated orbit, general relativity predicts that the long axis of the ellipse should rotate about the sun at a rate of about one degree in ten thousand years. Small though this effect is, it had been noticed before 1915 and served as one of the first confirmations of Einstein's theory. In recent years the even smaller deviations of the orbits of the other planets from the Newtonian predictions have been measured by radar and found to agree with the predictions of general relativity.

Light rays too must follow geodesics in space-time. Again, the fact that space is curved means that light no longer appears to travel in straight lines in space. So general relativity predicts that light should be bent by gravitational fields. For example, the theory predicts that the light cones of points near the sun would be slightly bent inward, on account of the mass of the sun. This means that light from a distant star that happened to pass near the sun would be deflected through a small angle, causing the star to appear in a different position to an observer on the earth (Fig. 2.9). Of course, if the light from the star always passed close to the sun, we would not be able to tell whether the light was being deflected or if instead the star was really where we see it. However, as the earth orbits around the sun, different stars appear to pass behind the sun and have their light deflected. They therefore change their apparent position relative to other stars.

It is normally very difficult to see this effect, because the light from the sun makes it impossible to observe stars that appear near to the sun in the sky. However, it is possible to do so during an eclipse of the sun, when the sun's light is blocked out by the moon. Einstein's prediction of light deflection could not be tested immediately in 1915, because the First World War was in progress, and it was not until 1919 that a British expedition, observing an eclipse from West

Africa, showed that light was indeed deflected by the sun, just as predicted by the theory. This proof of a German theory by British scientists was hailed as a great act of reconciliation between the two countries after the war. It is ironic, therefore, that later examination of the photographs taken on that expedition showed the errors were as great as the effect they were trying to measure. Their measurement had been sheer luck, or a case of knowing the result they wanted to get, not an uncommon occurrence in science. The light deflection has, however, been accurately confirmed by a number of later observations.

Another prediction of general relativity is that time should appear to run slower near a massive body like the earth. This is because there is a relation between the energy of light and its frequency (that is, the number of waves of light per second): the greater the energy, the higher the frequency. As light travels upward in the earth's gravitational field, it loses energy, and so its frequency goes down. (This means that the length of time between one wave crest and the next goes up.) To someone high up, it would appear that everything down below was taking longer to happen. This prediction was tested in 1962, using a pair of very accurate clocks mounted at the top and bottom of a water tower. The clock at the bottom, which was nearer the earth, was found to run slower, in exact agreement with general relativity. The difference in the speed of clocks at different heights above the earth is now of considerable practical importance, with the advent of very accurate navigation systems based on signals from satellites. If one ignored the predictions of general relativity, the position that one calculated would be wrong by several miles!

Newton's laws of motion put an end to the idea of absolute position in space. The theory of relativity gets rid of absolute time. Consider a pair of twins. Suppose that one twin goes to live on the top of a mountain while the other stays at sea level. The first twin would age faster than the second. Thus, if they met again, one would be older than the other. In this case, the difference in ages would be very small, but it would be much larger if one of the twins went for a long trip in a spaceship at nearly the speed of light. When he returned, he would be much younger than the one who stayed on Earth. This is known as the twins paradox, but it is a paradox only if one has the idea of absolute time at the back of one's mind. In the theory of relativity there is no unique absolute time, but instead each individual has his own personal measure of time that depends on where he is and how he is moving.

Before 1915, space and time were thought of as a fixed arena in which events took place, but which was not affected by what happened in it. This was true even of the special theory of relativity. Bodies moved, forces attracted and repelled, but time and space simply continued, unaffected. It was natural to think that space and time went on forever.

The situation, however, is quite different in the general theory of relativity. Space and time are now dynamic quantities: when a body moves, or a force acts, it affects the curvature of space and time—and in turn the structure of space-time affects the way in which bodies move and forces act. Space and time not only affect but also are affected by everything that happens in the universe. Just as one cannot talk about events in the universe without the notions of space and time, so in general relativity it became meaningless to talk about space and time outside the limits of the universe.

In the following decades this new understanding of space and time was to revolutionize our view of the universe. The old idea of an essentially unchanging universe that could have existed, and could continue to exist, forever was replaced by the notion of a dynamic, expanding universe that seemed to have begun a finite time ago, and that might end at a finite time in the future. That revolution forms the subject of the next chapter. And years later, it was also to be the starting point for my work in theoretical physics. Roger Penrose and I showed that Einstein's general theory of relativity implied that the universe must have a beginning and, possibly, an end.

3

The Expanding Universe

If one looks at the sky on a clear, moonless night, the brightest objects one sees are likely to be the planets Venus, Mars, Jupiter, and Saturn. There will also be a very large number of stars, which are just like our own sun but much farther from us. Some of these fixed stars do, in fact, appear to change very slightly their positions relative to each other as the earth orbits around the sun: they are not really fixed at all! This is because they are comparatively near to us. As the earth goes round the sun, we see them from different positions against the background of more distant stars. This is fortunate, because it enables us to measure directly the distance of these stars from us: the nearer they are, the more they appear to move. The nearest star, called Proxima Centauri, is found to be about four light-years away (the light from it takes about four years to reach Earth), or about twenty-three million million miles. Most of the other stars that are visible to the naked eye lie within a few hundred light-years of us. Our sun, for comparison, is a mere eight light-minutes away!

The visible stars appear spread all over the night sky, but are particularly concentrated in one band, which we call the Milky Way. As long ago as 1750, some astronomers were suggesting that the appearance of the Milky Way could be explained if most of the visible stars lie in a single disklike configuration, one example of what we now call a spiral galaxy. Only a few decades later, the astronomer, Sir William Herschel, confirmed his idea by painstakingly cataloging the positions and distances of vast numbers of stars. Even so, the idea gained complete acceptance only early this century.

Our modern picture of the universe dates back to only 1924, when the American astronomer Edwin Hubble demonstrated that ours was not the only galaxy. There were in fact many others, with vast tracts of empty space between them. In order to prove this, he needed to determine the distances to these other galaxies, which are so far away that, unlike nearby stars, they really do appear fixed. Hubble was forced, therefore, to use indirect methods to measure the distances. Now, the apparent brightness of a star depends on two factors: how much light it radiates (its luminosity), and how far it is from us. For nearby stars, we can measure their apparent brightness and their distance, and so we can work out their luminosity. Conversely, if we knew the luminosity of stars in other galaxies, we could work out their distance by measuring their apparent brightness. Hubble noted that certain types of stars always have the same luminosity when they are near enough for us to measure; therefore, he argued, if we found such stars in another galaxy, we could assume that they had the same luminosity—and so calculate the distance to that galaxy. If we could do this for a number of stars in the same galaxy, and our calculations always gave the same distance, we could be fairly confident of our estimate.

In this way, Edwin Hubble worked out the distances to nine different galaxies. We now know that our galaxy is only one of some hundred thousand million that can be seen using modern telescopes, each galaxy itself containing some hun-

FIGURE 3.1

dred thousand million stars. Fig. 3.1 shows a picture of one spiral galaxy that is similar to what we think ours must look like to someone living in another galaxy. We live in a galaxy that is about one hundred thousand light-years across and is slowly rotating; the stars in its spiral arms orbit around its center about once every several hundred million years. Our sun is just an ordinary, average-sized, yellow star, near the inner edge of one of the spiral arms. We have certainly come a long way since Aristotle and Ptolemy, when we thought that the earth was the center of the universe!

Stars are so far away that they appear to us to be just pinpoints of light. We cannot see their size or shape. So how can we tell different types of stars apart? For the vast majority of stars, there is only one characteristic feature that we can observe—the color of their light. Newton discovered that if light from the sun passes through a triangular-shaped piece of glass, called a prism, it breaks up into its component colors (its spectrum) as in a rainbow. By focusing a telescope on an individual star or galaxy, one can similarly observe the spectrum of the light from that star or galaxy. Different stars have different spectra, but the relative brightness of the different colors is always exactly what one would expect to find in the light emitted by an object that is glowing red hot. (In fact, the light emitted by any opaque object that is glowing red hot has a characteristic spectrum that depends only on its temperature—a thermal spectrum. This means that we can tell a star's temperature from the spectrum of its light.) Moreover, we find that certain very specific colors are missing from stars' spectra, and these missing colors may vary from star to star. Since we know that each chemical element absorbs a characteristic set of very specific colors, by matching these to those that are missing from a star's spectrum, we can determine exactly which elements are present in the star's atmosphere.

In the 1920s, when astronomers began to look at the spectra of stars in other galaxies, they found something most peculiar: there were the same characteristic sets of missing colors as for stars in our own galaxy, but they were all shifted by the same relative amount toward the red end of the spectrum. To understand the implications of this, we must first understand the Doppler effect. As we have seen, visible light consists of fluctuations, or waves, in the electromagnetic field. The frequency (or number of waves per second) of light is extremely high, ranging from four to seven hundred million million waves per second. The different frequencies of light are what the human eye sees as different colors, with the lowest frequencies appearing at the red end of the spectrum and the highest frequencies at the blue end. Now imagine a source of light at a constant distance from us, such as a star, emitting waves of light at a constant frequency. Obviously the frequency of the waves we receive will be the same as the frequency at which they are emitted (the gravitational field of the galaxy will not be large enough to have a significant effect). Suppose now that the source starts moving toward us. When the source emits the next wave crest it will be nearer to us, so the time that wave crest takes to reach us will be less than when the star was stationary. This means that the time between the two wave crests reaching us is smaller, and therefore the number of waves we receive each second (i.e., the frequency) is higher than when the star was stationary. Correspondingly, if the source is moving away from us, the frequency of the waves we receive will be lower. In the case of light, therefore, this means that stars moving away from us will have their spectra shifted toward the red end of the spectrum (red-shifted) and those moving toward us will have their spectra blue-shifted. This relationship between frequency and speed, which is called the Doppler effect, is an everyday experience. Listen to a car passing on the road: as the car is approaching, its engine sounds at a higher pitch (corresponding to a higher frequency of sound waves), and when it passes and goes away, it sounds at a lower pitch. The behavior of light or radio waves is similar. Indeed, the police make use of the Doppler effect to measure the speed of cars by measuring the frequency of pulses of radio waves reflected off them.

In the years following his proof of the existence of other galaxies, Hubble spent his time cataloging their distances and observing their spectra. At that time most people expected the galaxies to be moving around quite randomly, and so expected to find as many blue-shifted spectra as red-shifted ones. It was quite a surprise, therefore, to find that most galaxies appeared red-shifted: nearly all were moving away from us! More surprising still was the finding that Hubble published in 1929: even the size of a galaxy's red shift is not random, but is directly proportional to the galaxy's distance from us. Or, in other words, the farther a galaxy is, the faster it is moving away! And that meant that the universe could not be static, as everyone previously had thought, but is in fact expanding; the distance between the different galaxies is growing all the time.

The discovery that the universe is expanding was one of the great intellectual revolutions of the twentieth century. With hindsight, it is easy to wonder why no one had thought of it before. Newton, and others, should have realized that a static universe would soon start to contract under the influence of gravity. But suppose instead the universe expanding. If it was expanding fairly slowly, the force of gravity would cause it eventually to stop expanding and then to start contracting. However, if it was expanding at more than a certain critical rate, gravity would never be strong enough to stop it, and the universe would continue to expand forever. This is a bit like what happens when one fires a rocket upward from the surface of the earth. If it has a fairly low speed, gravity will eventually stop the rocket and it will start falling back. On the other hand, if the rocket has more than a certain critical speed (about seven miles per second) gravity will not be strong enough to pull it back, so it will keep going away from the earth forever. This behavior of the universe could have been predicted from Newton's theory of gravity at any time in the nineteenth, the eighteenth, or even the late seventeenth centuries. Yet so strong was the belief in a static universe that it persisted into the early twentieth century. Even Einstein, when he formulated the general theory of relativity in 1915, was so sure that the universe had to be static that he modified his theory to make this possible, introducing a so-called cosmological constant into his equations. Einstein introduced a new "antigravity" force, which, unlike other forces, did not come from any particular source, but was built into the very fabric of space-time. He claimed that space-time had an inbuilt tendency to expand, and this could be made to balance exactly the attraction of all the matter in the universe, so that a static universe would result. Only one man, it seems, was willing to take general relativity at face value, and while Einstein and other physicists were looking for ways of avoiding general relativity's prediction of a nonstatic universe, the Russian physicist and mathematician Alexander Friedmann instead set about explaining it.

Friedmann made two very simple assumptions about the universe: that the universe looks identical in whichever direction we look, and that this would also be true if we were observing the universe from anywhere else. From these two ideas alone, Friedmann showed that we should not expect the universe to be static. In fact, in 1922, several years before Edwin Hubble's discovery, Friedmann predicted exactly what Hubble found!

The assumption that the universe looks the same in every direction is clearly not true in reality. For example, as we have seen, the other stars in our galaxy form a distinct band of light across the night sky, called the Milky Way. But if we look at distant galaxies, there seems to be more or less the same number of them. So the universe does seem to be roughly the same in every direction, provided one views it on a scale large compared to the distance between galaxies, and ignores the differences on small scales. For a long time, this was sufficient justification for Friedmann's assumption—as a rough approximation to the real universe. But more recently a lucky accident uncovered the fact that Friedmann's assumption is in fact a remarkably accurate description of our universe.

In 1965 two American physicists at the Bell Telephone Laboratories in New Jersey, Arno Penzias and Robert Wilson, were testing a very sensitive microwave detector. (Microwaves are just like light waves, but with a frequency of the order of only ten thousand million waves per second.) Penzias and Wilson were worried when they found that their detector was picking up more noise than it ought to. The noise did not appear to be coming from any particular direction. First they discovered bird droppings in their detector and checked for other possible malfunctions, but soon ruled these out. They knew that any noise from within the atmosphere would be stronger when the detector was not pointing straight up than when it was, because light rays travel through much more atmosphere when received from near the horizon than when received from directly overhead. The extra noise was the same whichever direction the detector was pointed, so it must come from *outside* the atmosphere. It was also the same day and night and throughout the year, even though the earth was rotating on its axis and orbiting around the sun. This showed that the radiation must come from beyond the Solar System, and even from beyond the galaxy, as otherwise it would vary as the movement of Earth pointed the detector in different directions. In fact, we know that the radiation must have traveled to us across most of the observable universe, and since it appears to be the same in different directions, the universe must also be the same in every direction, if only on a large scale. We now know that whichever direction we look, this noise never varies by more than one part in ten thousand—so Penzias and Wilson had unwittingly stumbled across a remarkably accurate confirmation of Friedmann's first assumption.

At roughly the same time two American physicists at nearby Princeton University, Bob Dicke and Jim Peebles, were also taking an interest in microwaves. They were working on a suggestion, made by George Gamow (once a student of Alexander Friedmann), that the early universe should have been very hot and dense, glowing white hot. Dicke and Peebles argued that we should still be able to see the glow of the early universe, because light from very distant parts of it would only just be reaching us now. However, the expansion of the universe meant that this light should be so greatly red-shifted that it would appear to us now as microwave radiation. Dicke and Peebles were preparing to look for this radiation when Penzias and Wilson heard about their work and realized that they had already found it. For this, Penzias and Wilson were awarded the Nobel prize in 1978 (which seems a bit hard on Dicke and Peebles, not to mention Gamow!).

Now at first sight, all this evidence that the universe looks the same whichever direction we look in might seem to suggest there is something special about our place in the universe. In particular, it might seem that if we observe all other galaxies to be moving away from us, then we must be at the center of the universe. There is, however, an alternate explanation: the universe might look the same in every direction as seen from any other galaxy, too. This, as we have seen, was Friedmann's second assumption. We have no scientific evidence for, or against, this assumption. We believe it only on grounds of modesty: it would be most remarkable if the universe looked the same in every direction around us, but not around other points in the universe! In Friedmann's model, all the galaxies are moving directly away from each other. The situation is rather like a balloon with a number of spots painted on it being steadily blown up. As the balloon expands, the distance between any two spots increases, but there is no spot that can be said to be the center of the expansion. Moreover, the farther apart the spots are, the faster they will be moving apart. Similarly, in Friedmann's model the speed at which any two galaxies are moving apart is proportional to the distance between them. So it predicted that the red shift of a galaxy should be directly proportional to its distance from us, exactly as Hubble found. Despite the success of his model and his prediction of Hubble's observations, Friedmann's work remained largely unknown in the West until similar models were discovered in 1935 by the American physicist Howard Robertson and the British mathematician Arthur Walker, in response to Hubble's discovery of the uniform expansion of the universe.

Although Friedmann found only one, there are in fact three different kinds of models that obey Friedmann's two fundamental assumptions. In the first kind (which Friedmann found) the universe is expanding sufficiently slowly that the gravita-

FIGURE 3.2

FIGURE 3.3

FIGURE 3.4

tional attraction between the different galaxies causes the expansion to slow down and eventually to stop. The galaxies then start to move toward each other and the universe contracts. Fig. 3.2 shows how the distance between two neighboring galaxies changes as time increases. It starts at zero, increases to a maximum, and then decreases to zero again. In the second kind of solution, the universe is expanding so rapidly that the gravitational attraction can never stop it, though it does slow it down a bit. Fig. 3.3 shows the separation between neighboring galaxies in this model. It starts at zero and eventually the galaxies are moving apart at a steady speed. Finally, there is a third kind of solution, in which the universe is expanding only just fast enough to avoid recollapse. In this case the separation, shown in Fig. 3.4, also starts at zero and increases forever. However, the speed at which the galaxies are moving apart gets smaller and smaller, although it never quite reaches zero.

A remarkable feature of the first kind of Friedmann model

is that in it the universe is not infinite in space, but neither does space have any boundary. Gravity is so strong that space is bent round onto itself, making it rather like the surface of the earth. If one keeps traveling in a certain direction on the surface of the earth, one never comes up against an impassable barrier or falls over the edge, but eventually comes back to where one started. In the first Friedmann model, space is just like this, but with three dimensions instead of two for the earth's surface. The fourth dimension, time, is also finite in extent, but it is like a line with two ends or boundaries, a beginning and an end. We shall see later that when one combines general relativity with the uncertainty principle of quantum mechanics, it is possible for both space and time to be finite without any edges or boundaries.

The idea that one could go right round the universe and end up where one started makes good science fiction, but it doesn't have much practical significance, because it can be shown that the universe would recollapse to zero size before one could get round. You would need to travel faster than light in order to end up where you started before the universe came to an end—and that is not allowed!

In the first kind of Friedmann model, which expands and recollapses, space is bent in on itself, like the surface of the earth. It is therefore finite in extent. In the second kind of model, which expands forever, space is bent the other way, like the surface of a saddle. So in this case space is infinite. Finally, in the third kind of Friedmann model, with just the critical rate of expansion, space is flat (and therefore is also infinite).

But which Friedmann model describes our universe? Will the universe eventually stop expanding and start contracting, or will it expand forever? To answer this question we need to know the present rate of expansion of the universe and its present average density. If the density is less than a certain critical value, determined by the rate of expansion, the gravitational attraction will be too weak to halt the expansion. If the density is greater than the critical value, gravity will stop the expansion at some time in the future and cause the universe to recollapse.

We can determine the present rate of expansion by measuring the velocities at which other galaxies are moving away from us, using the Doppler effect. This can be done very accurately. However, the distances to the galaxies are not very well known because we can only measure them indirectly. So all we know is that the universe is expanding by between 5 percent and 10 percent every thousand million years. However, our uncertainty about the present average density of the universe is even greater. If we add up the masses of all the stars that we can see in our galaxy and other galaxies, the total is less than one hundredth of the amount required to halt the expansion of the universe, even for the lowest estimate of the rate of expansion. Our galaxy and other galaxies, however, must contain a large amount of "dark matter" that we cannot see directly, but which we know must be there because of the influence of its gravitational attraction on the orbits of stars in the galaxies. Moreover, most galaxies are found in clusters, and we can similarly infer the presence of yet more dark matter in between the galaxies in these clusters by its effect on the motion of the galaxies. When we add up all this dark matter, we still get only about one tenth of the amount required to halt the expansion. However, we cannot exclude the possibility that there might be some other form of matter, distributed almost uniformly throughout the universe, that we have not yet detected and that might still raise the average density of the universe up to the critical value needed to halt the expansion. The present evidence therefore suggests that the universe will probably expand forever, but all we can really be sure of is that even if the universe is going to recollapse, it won't do so for at least another ten thousand million years, since it has already been expanding for at least that long. This should not unduly worry us: by that time, unless we have colonized beyond the Solar System, mankind will long since have died out, extinguished along with our sun!

All of the Friedmann solutions have the feature that at some time in the past (between ten and twenty thousand million years ago) the distance between neighboring galaxies must have been zero. At that time, which we call the big bang, the density of the universe and the curvature of space-time would have been infinite. Because mathematics cannot really handle infinite numbers, this means that the general theory of relativity (on which Friedmann's solutions are based) predicts that there is a point in the universe where the theory itself breaks down. Such a point is an example of what mathematicians call a singularity. In fact, all our theories of science are formulated on the assumption that space-time is smooth and nearly flat, so they break down at the big bang singularity, where the curvature of space-time is infinite. This means that even if there were events before the big bang, one could not use them to determine what would happen afterward, because predictability would break down at the big bang. Correspondingly, if, as is the case, we know only what has happened since the big bang, we could not determine what happened beforehand. As far as we are concerned, events before the big bang can have no consequences, so they should not form part of a scientific model of the universe. We should therefore cut them out of the model and say that time had a beginning at the big bang.

Many people do not like the idea that time has a beginning, probably because it smacks of divine intervention. (The Catholic Church, on the other hand, seized on the big bang model and in 1951 officially pronounced it to be in accordance with the Bible.) There were therefore a number of attempts to avoid the conclusion that there had been a big bang. The proposal that gained widest support was called the steady state theory. It was suggested in 1948 by two refugees from Nazi-occupied Austria, Hermann Bondi and Thomas Gold, together with a Briton, Fred Hoyle, who had worked with them on the development of radar during the war. The idea was that as the galaxies moved away from each other, new galaxies were continually forming in the gaps in between, from new matter that was being continually created. The universe would therefore look roughly the same at all times as well as at all points of space. The steady state theory required a modification of general relativity to allow for the continual creation of matter, but the rate that was involved was so low (about one particle per cubic kilometer per year) that it was not in conflict with experiment. The theory was a good scientific theory, in the sense described in Chapter 1: it was simple and it made definite predictions that could be tested by observation. One of these predictions was that the number of galaxies or similar objects in any given volume of space should be the same wherever and whenever we look in the universe. In the late 1950s and early 1960s a survey of sources of radio waves from outer space was carried out at Cambridge by a group of astronomers led by Martin Ryle (who had also worked with Bondi, Gold, and Hoyle on radar during the war). The Cambridge group showed that most of these radio sources must lie outside our galaxy (indeed many of them could be identified with other galaxies) and also that there were many more weak sources than strong ones. They interpreted the weak sources as being the more distant ones, and the stronger ones as being nearer. Then there appeared to be less common sources per unit volume of space for the nearby sources than for the distant ones. This could mean that we are at the center of a great region in the universe in which the sources are fewer than elsewhere. Alternatively, it could mean that the sources were more numerous in the past, at the time that the radio waves left on their journey to us, than they are now. Either explanation contradicted the predictions of the steady state theory. Moreover, the discovery of the microwave radiation by Penzias and Wilson in 1965 also indicated that the universe must have been much denser in the past. The steady state theory therefore had to be abandoned.

Another attempt to avoid the conclusion that there must

have been a big bang, and therefore a beginning of time, was made by two Russian scientists, Evgenii Lifshitz and Isaac Khalatnikov, in 1963. They suggested that the big bang might be a peculiarity of Friedmann's models alone, which after all were only approximations to the real universe. Perhaps, of all the models that were roughly like the real universe, only Friedmann's would contain a big bang singularity. In Friedmann's models, the galaxies are all moving directly away from each other—so it is not surprising that at some time in the past they were all at the same place. In the real universe, however, the galaxies are not just moving directly away from each other—they also have small sideways velocities. So in reality they need never have been all at exactly the same place, only very close together. Perhaps then the current expanding universe resulted not from a big bang singularity, but from an earlier contracting phase; as the universe had collapsed the particles in it might not have all collided, but had flown past and then away from each other, producing the present expansion of the universe. How then could we tell whether the real universe should have started out with a big bang? What Lifshitz and Khalatnikov did was to study models of the universe that were roughly like Friedmann's models but took account of the irregularities and random velocities of galaxies in the real universe. They showed that such models could start with a big bang, even though the galaxies were no longer always moving directly away from each other, but they claimed that this was still only possible in certain exceptional models in which the galaxies were all moving in just the right way. They argued that since there seemed to be infinitely more Friedmann-like models without a big bang singularity than there were with one, we should conclude that there had not in reality been a big bang. They later realized, however, that there was a much more general class of Friedmann-like models that did have singularities, and in which the galaxies did not have to be moving any special way. They therefore withdrew their claim in 1970.

The work of Lifshitz and Khalatnikov was valuable because it showed that the universe *could* have had a singularity, a big bang, if the general theory of relativity was correct. However, it did not resolve the crucial question: Does general relativity predict that our universe *should* have had a big bang, a beginning of time? The answer to this came out of a completely different approach introduced by a British mathematician and physicist, Roger Penrose, in 1965. Using the way light cones behave in general relativity together with the fact that gravity is always attractive, he showed that a star collapsing under its own gravity is trapped in a region whose surface eventually shrinks to zero size. And, since the surface of the region shrinks to zero, so too must its volume. All the matter in the star will be compressed into a region of zero volume, so the density of matter and the curvature of space-time become infinite. In other words, one has a singularity contained within a region of space-time known as a black hole.

At first sight, Penrose's result applied only to stars; it didn't have anything to say about the question of whether the entire universe had a big bang singularity in its past. However, at the time that Penrose produced his theorem, I was a research student desperately looking for a problem with which to complete my Ph.D. thesis. Two years before I had been diagnosed as suffering from ALS, commonly known as Lou Gehrig's disease, or motor neuron disease, and given to understand that I had only one or two more years to live. In these circumstances there had not seemed much point in working on my Ph.D.—I did not expect to survive that long. Yet two years had gone by and I was not that much worse. In fact, things were going rather well for me and I had gotten engaged to a very nice girl, Jane Wilde. But in order to get married, I needed a job, and in order to get a job, I needed a Ph.D.

In 1965 I read about Penrose's theorem that any body undergoing gravitational collapse must eventually form a singularity. I soon realized that if one reversed the direction of time in Penrose's theorem, so that the collapse became an expansion, the conditions of his theorem would still hold, provided the universe were roughly like a Friedmann model on large scales at the present time. Penrose's theorem had shown that any collapsing star *must* end in a singularity; the time-reversed argument showed that any Friedmann-like expanding universe *must* have begun with a singularity. For technical reasons, Penrose's theorem required that the universe be infinite in space. So I could in fact use it to prove that there should be a singularity only if the universe was expanding fast enough to avoid collapsing again (since only those Friedmann models were infinite in space).

During the next few years I developed new mathematical techniques to remove this and other technical conditions from the theorems that proved that singularities must occur. The final result was a joint paper by Penrose and myself in 1970, which at last proved that there must have been a big bang singularity provided only that general relativity is correct and the universe contains as much matter as we observe. There was a lot of opposition to our work, partly from the Russians because of their Marxist belief in scientific determinism, and partly from people who felt that the whole idea of singularities was repugnant and spoiled the beauty of Einstein's theory. However, one cannot really argue with a mathematical theorem. So in the end our work became generally accepted and nowadays nearly everyone assumes that the universe started with a big bang singularity. It is perhaps ironic that, having changed my mind, I am now trying to convince other physicists that there was in fact no singularity at the beginning of the universe—as we shall see later, it can disappear once quantum effects are taken into account.

We have seen in this chapter how, in less than half a century, man's view of the universe, formed over millennia, has been transformed. Hubble's discovery that the universe was expanding, and the realization of the insignificance of our own planet in the vastness of the universe, were just the starting point. As experimental and theoretical evidence mounted, it became more and more clear that the universe must have had a beginning in time, until in 1970 this was finally proved by Penrose and myself, on the basis of Einstein's general theory of relativity. That proof showed that general relativity is only an incomplete theory: it cannot tell us how the universe started off, because it predicts that all physical theories, including itself, break down at the beginning of the universe. However, general relativity claims to be only a partial theory, so what the singularity theorems really show is that there must have been a time in the very early universe when the universe was so small, that one could no longer ignore the small-scale effects of the other great partial theory of the twentieth century, quantum mechanics. At the start of the 1970s, then, we were forced to turn our search for an understanding of the universe from our theory of the extraordinarily vast to our theory of the extraordinarily tiny. That theory, quantum mechanics, will be described next, before we turn to the efforts to combine the two partial theories into a single quantum theory of gravity.

4

The Uncertainty Principle

The success of scientific theories, particularly Newton's theory of gravity, led the French scientist the Marquis de Laplace at the beginning of the nineteenth century to argue that the universe was completely deterministic. Laplace suggested that there should be a set of scientific laws that would allow us to predict everything that would happen in the universe, if only we knew the complete state of the universe at one time. For example, if we knew the positions and speeds of the sun and the planets at one time, then we could use Newton's laws to calculate the state of the Solar System at any other time. Determinism seems fairly obvious in this case, but Laplace went further to assume that there were similar laws governing everything else, including human behavior.

The doctrine of scientific determinism was strongly resisted by many people, who felt that it infringed God's freedom to intervene in the world, but it remained the standard assumption of science until the early years of this century. One of the first indications that this belief would have to be abandoned came when calculations by the British scientists Lord Rayleigh and Sir James Jeans suggested that a hot object, or body, such as a star, must radiate energy at an infinite rate. According to the laws we believed at the time, a hot body ought to give off electromagnetic waves (such as radio waves, visible light, or X rays) equally at all frequencies. For example, a hot body should radiate the same amount of energy in waves with frequencies between one and two million million waves a second as in waves with frequencies between two and three million million waves a second. Now since the number of waves a second is unlimited, this would mean that the total energy radiated would be infinite.

In order to avoid this obviously ridiculous result, the German scientist Max Planck suggested in 1900 that light, X rays, and other waves could not be emitted at an arbitrary rate, but only in certain packets that he called quanta. Moreover, each quantum had a certain amount of energy that was greater the higher the frequency of the waves, so at a high enough frequency the emission of a single quantum would require more energy than was available. Thus the radiation at high frequencies would be reduced, and so the rate at which the body lost energy would be finite.

The quantum hypothesis explained the observed rate of emission of radiation from hot bodies very well, but its implications for determinism were not realized until 1926, when another German scientist, Werner Heisenberg, formulated his famous uncertainty principle. In order to predict the future position and velocity of a particle, one has to be able to measure its present position and velocity accurately. The obvious way to do this is to shine light on the particle. Some of the waves of light will be scattered by the particle and this will indicate its position. However, one will not be able to determine the position of the particle more accurately than the distance between the wave crests of light, so one needs to use light of a short wavelength in order to measure the position of the particle precisely. Now, by Planck's quantum hypothesis, one cannot use an arbitrarily small amount of light; one has to use at least one quantum. This quantum will disturb the particle and change its velocity in a way that cannot be predicted. Moreover, the more accurately one measures the position, the shorter the wavelength of the light that one needs and hence the higher the energy of a single quantum. So the velocity of the particle will be disturbed by a larger amount. In other words, the more accurately you try to measure the position of the particle, the less accurately you can measure its speed, and vice versa. Heisenberg showed that the uncertainty in the position of the particle times the uncertainty in its velocity times the mass of the particle can never be smaller than a certain quantity, which is known as Planck's constant. Moreover, this limit does not depend on the way in which one tries to measure the position or velocity of the particle, or on the type of particle: Heisenberg's uncertainty principle is a fundamental, inescapable property of the world.

The uncertainty principle had profound implications for the way in which we view the world. Even after more than fifty years they have not been fully appreciated by many philosophers, and are still the subject of much controversy. The uncertainty principle signaled an end to Laplace's dream of a theory of science, a model of the universe that would be completely deterministic: one certainly cannot predict future events exactly if one cannot even measure the present state of the universe precisely! We could still imagine that there is a set of laws that determines events completely for some supernatural being, who could observe the present state of the universe without disturbing it. However, such models of the universe are not of much interest to us ordinary mortals. It seems better to employ the principle of economy known as Occam's razor and cut out all the features of the theory that cannot be observed. This approach led Heisenberg, Erwin Schrödinger, and Paul Dirac in the 1920s to reformulate mechanics into a new theory called quantum mechanics, based on the uncertainty principle. In this theory particles no longer had separate, well-defined positions and velocities that could not be observed. Instead, they had a quantum state, which was a combination of position and velocity.

In general, quantum mechanics does not predict a single definite result for an observation. Instead, it predicts a number of different possible outcomes and tells us how likely each of these is. That is to say, if one made the same measurement on a large number of similar systems, each of which started off in the same way, one would find that the result of the measurement would be A in a certain number of cases, B in a different number, and so on. One could predict the approximate number of times that the result would be A or B, but one could not predict the specific result of an individual measurement. Quantum mechanics therefore introduces an unavoidable element of unpredictability or randomness into science. Einstein objected to this very strongly, despite the important role he had played in the development of these ideas. Einstein was awarded the Nobel prize for his contribution to quantum theory. Nevertheless, Einstein never accepted that the universe was governed by chance; his feelings were summed up in his famous statement "God does not play dice." Most other scientists, however, were willing to accept quantum mechanics because it agreed perfectly with experiment. Indeed, it has been an outstandingly successful theory and underlies nearly all of modern science and technology. It governs the behavior of transistors and integrated circuits, which are the essential components of electronic devices such as televisions and computers, and is also the basis of modern chemistry and biology. The only areas of physical science into which quantum mechanics has not yet been properly incorporated are gravity and the large-scale structure of the universe.

Although light is made up of waves, Planck's quantum hypothesis tells us that in some ways it behaves as if it were composed of particles: it can be emitted or absorbed only in packets, or quanta. Equally, Heisenberg's uncertainty principle implies that particles behave in some respects like waves: they do not have a definite position but are "smeared out" with a certain probability distribution. The theory of quantum mechanics is based on an entirely new type of mathematics that no longer describes the real world in terms of particles and waves; it is only the observations of the world that may be described in those terms. There is thus a duality between waves and particles in quantum mechanics: for some purposes it is helpful to think of particles as waves and for other purposes it is better to think of waves as particles. An important consequence of this is that one can observe what is called interference between two sets of waves or particles. That is to say, the crests of one set of waves may coincide with the troughs of the other set. The two sets of waves then cancel

IN PHASE

WAVE CRESTS AND TROUGHS REINFORCE EACH OTHER

OUT OF PHASE

WAVE CRESTS AND TROUGHS CANCEL OUT

FIGURE 4.1

each other out, rather than adding up to a stronger wave as one might expect (Fig. 4.1). A familiar example of interference in the case of light is the colors that are often seen in soap bubbles. These are caused by reflection of light from the two sides of the thin film of water forming the bubble. White light consists of light waves of all different wavelengths, or colors. For certain wavelengths the crests of the waves reflected from one side of the soap film coincide with the troughs reflected from the other side. The colors correspond-

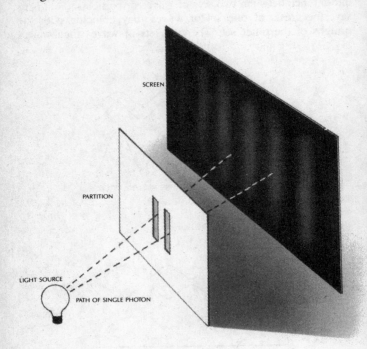

FIGURE 4.2

ing to these wavelengths are absent from the reflected light, which therefore appears to be colored.

Interference can also occur for particles, because of the duality introduced by quantum mechanics. A famous example is the so-called two-slit experiment (Fig. 4.2). Consider a partition with two narrow parallel slits in it. On one side of the partition one places a source of light of a particular color (that is, of a particular wavelength). Most of the light will hit the partition, but a small amount will go through the slits. Now suppose one places a screen on the far side of the partition from the light. Any point on the screen will receive waves from the two slits. However, in general, the distance the light has to travel from the source to the screen via the two slits will be different. This will mean that the waves from the slits will not be in phase with each other when they arrive at the screen: in some places the waves will cancel each other out, and in others they will reinforce each other. The result is a characteristic pattern of light and dark fringes.

The remarkable thing is that one gets exactly the same kind of fringes if one replaces the source of light by a source of particles such as electrons with a definite speed (this means that the corresponding waves have a definite length). It seems the more peculiar because if one only has one slit, one does not get any fringes, just a uniform distribution of electrons across the screen. One might therefore think that opening another slit would just increase the number of electrons hitting each point of the screen, but, because of interference, it actually decreases it in some places. If electrons are sent through the slits one at a time, one would expect each to pass through one slit or the other, and so behave just as if the slit it passed through were the only one there—giving a uniform distribution on the screen. In reality, however, even when the electrons are sent one at a time, the fringes still appear. Each electron, therefore, must be passing through *both* slits at the same time!

The phenomenon of interference between particles has been crucial to our understanding of the structure of atoms, the basic units of chemistry and biology and the building blocks out of which we, and everything around us, are made. At the beginning of this century it was thought that atoms were rather like the planets orbiting the sun, with electrons (particles of negative electricity) orbiting around a central nucleus, which carried positive electricity. The attraction between the positive and negative electricity was supposed to keep the electrons in their orbits in the same way that the gravitational attraction between the sun and the planets keeps the planets in their orbits. The trouble with this was that the laws of mechanics and electricity, before quantum mechanics, predicted that the electrons would lose energy and so spiral inward until they collided with the nucleus. This would mean that the atom, and indeed all matter, should rapidly collapse to a state of very high density. A partial solution to this problem was found by the Danish scientist Niels Bohr in 1913. He suggested that maybe the electrons were not able to orbit at just any distance from the central nucleus but only at certain specified distances. If one also supposed that only one or two electrons could orbit at any one of these distances, this would solve the problem of the collapse of the atom, because the electrons could not spiral in any farther than to fill up the orbits with the least distances and energies.

This model explained quite well the structure of the simplest atom, hydrogen, which has only one electron orbiting around the nucleus. But it was not clear how one ought to extend it to more complicated atoms. Moreover, the idea of a limited set of allowed orbits seemed very arbitrary. The new theory of quantum mechanics resolved this difficulty. It revealed that an electron orbiting around the nucleus could be thought of as a wave, with a wavelength that depended on its velocity. For certain orbits, the length of the orbit would correspond to a whole number (as opposed to a fractional number) of wavelengths of the electron. For these orbits the wave crest would be in the same position each time round, so the waves would add up: these orbits would correspond to Bohr's allowed orbits. However, for orbits whose lengths were not a whole number of wavelengths, each wave crest would eventually be canceled out by a trough as the electrons went round; these orbits would not be allowed.

A nice way of visualizing the wave/particle duality is the so-called sum over histories introduced by the American scientist Richard Feynman. In this approach the particle is not supposed to have a single history or path in space-time, as it would in a classical, nonquantum theory. Instead it is supposed to go from A to B by every possible path. With each path there are associated a couple of numbers: one represents the size of a wave and the other represents the position in the cycle (i.e., whether it is at a crest or a trough). The probability of going from A to B is found by adding up the waves for all the paths. In general, if one compares a set of neighboring paths, the phases or positions in the cycle will differ greatly. This means that the waves associated with these paths will almost exactly cancel each other out. However, for some sets of neighboring paths the phase will not vary much between paths. The waves for these paths will not cancel out. Such paths correspond to Bohr's allowed orbits.

With these ideas, in concrete mathematical form, it was relatively straightforward to calculate the allowed orbits in more complicated atoms and even in molecules, which are made up of a number of atoms held together by electrons in orbits that go round more than one nucleus. Since the structure of molecules and their reactions with each other underlie all of chemistry and biology, quantum mechanics allows us in principle to predict nearly everything we see around us, within the limits set by the uncertainty principle. (In practice, however, the calculations required for systems containing more than a few electrons are so complicated that we cannot do them.)

Einstein's general theory of relativity seems to govern the large-scale structure of the universe. It is what is called a classical theory; that is, it does not take account of the uncertainty principle of quantum mechanics, as it should for consistency with other theories. The reason that this does not lead to any discrepancy with observation is that all the gravitational fields that we normally experience are very weak. However, the singularity theorems discussed earlier indicate that the gravitational field should get very strong in at least two situations, black holes and the big bang. In such strong fields the effects of quantum mechanics should be important. Thus, in a sense, classical general relativity, by predicting points of infinite density, predicts its own downfall, just as classical (that is, nonquantum) mechanics predicted its downfall by suggesting that atoms should collapse to infinite density. We do not yet have a complete consistent theory that unifies general relativity and quantum mechanics, but we do know a number of the features it should have. The consequences that these would have for black holes and the big bang will be described in later chapters. For the moment, however, we shall turn to the recent attempts to bring together our understanding of the other forces of nature into a single, unified quantum theory.

5

Elementary
Particles and the
Forces of Nature

Aristotle believed that all the matter in the universe was made up of four basic elements, earth, air, fire, and water. These elements were acted on by two forces: gravity, the tendency for earth and water to sink, and levity, the tendency for air and fire to rise. This division of the contents of the universe into matter and forces is still used today.

Aristotle believed that matter was continuous, that is, one could divide a piece of matter into smaller and smaller bits without any limit: one never came up against a grain of matter that could not be divided further. A few Greeks, however, such as Democritus, held that matter was inherently grainy and that everything was made up of large numbers of various different kinds of atoms. (The word *atom* means "indivisible" in Greek.) For centuries the argument continued without any real evidence on either side, but in 1803 the British chemist and physicist, John Dalton, pointed out that the fact that chemical compounds always combined in certain proportions could be explained by the grouping together of atoms to form units called molecules. However, the argument between the two schools of thought was not finally settled in favor of the atomists until the early years of this century. One of the important pieces of physical evidence was provided by Einstein. In a paper written in 1905, a few weeks before the famous paper on special relativity, Einstein pointed out that what was called Brownian motion—the irregular, random motion of small particles of dust suspended in a liquid—could be explained as the effect of atoms of the liquid colliding with the dust particles.

By this time there were already suspicions that these atoms were not, after all, indivisible. Several years previously a fellow of Trinity College, Cambridge, J. J. Thomson, had demonstrated the existence of a particle of matter, called the electron, that had a mass less than one thousandth of that of the lightest atom. He used a setup rather like a modern TV picture tube: a red-hot metal filament gave off the electrons, and because these have a negative electric charge, an electric field could be used to accelerate them toward a phosphor-coated screen. When they hit the screen, flashes of light were generated. Soon it was realized that these electrons must be coming from within the atoms themselves, and in 1911 the British physicist Ernest Rutherford finally showed that the atoms of matter do have internal structure: they are made up of an extremely tiny, positively charged nucleus, around which a number of electrons orbit. He deduced this by analyzing the way in which α-particles, which are positively charged particles given off by radioactive atoms, are deflected when they collide with atoms.

At first it was thought that the nucleus of the atom was made up of electrons and different numbers of a positively charged particle called the proton, from the Greek word meaning "first," because it was believed to be the fundamental unit from which matter was made. However, in 1932 a colleague of Rutherford's at Cambridge, James Chadwick, discovered that the nucleus contained another particle, called the neutron, which had almost the same mass as a proton but no electrical charge. Chadwick received the Nobel prize for his discovery, and was elected Master of Gonville and Caius College, Cambridge (the college of which I am now a fellow). He later resigned as master because of disagreements with the fellows. There had been a bitter dispute in the college ever since a group of young fellows returning after the war had voted many of the old fellows out of the college offices they had held for a long time. This was before my time; I joined the college in 1965 at the tail end of the bitterness, when similar disagreements forced another Nobel-prize-winning master, Sir Nevill Mott, to resign.

Up to about twenty years ago, it was thought that protons and neutrons were "elementary" particles, but experiments in which protons were collided with other protons or electrons at high speeds indicated that they were in fact made up of smaller particles. These particles were named quarks by the Caltech physicist Murray Gell-Mann, who won the Nobel prize in 1969 for his work on them. The origin of the name is an enigmatic quotation from James Joyce: "Three quarks for Muster Mark!" The word *quark* is supposed to be pronounced like *quart,* but with a *k* at the end instead of a *t,* but is usually pronounced to rhyme with *lark.*

There are a number of different varieties of quarks: there are thought to be at least six "flavors," which we call up, down, strange, charmed, bottom, and top. Each flavor comes in three "colors," red, green, and blue. (It should be emphasized that these terms are just labels: quarks are much smaller than the wavelength of visible light and so do not have any color in the normal sense. It is just that modern physicists seem to have more imaginative ways of naming new particles and phenomena—they no longer restrict themselves to Greek!) A proton or neutron is made up of three quarks, one of each color. A proton contains two up quarks and one down quark; a neutron contains two down and one up. We can create particles made up of the other quarks (strange, charmed, bottom, and top), but these all have a much greater mass and decay very rapidly into protons and neutrons.

We now know that neither the atoms nor the protons and neutrons within them are indivisible. So the question is: What are the truly elementary particles, the basic building blocks from which everything is made? Since the wavelength of light is much larger than the size of an atom, we cannot hope to "look" at the parts of an atom in the ordinary way. We need to use something with a much smaller wavelength. As we saw in the last chapter, quantum mechanics tells us that all particles are in fact waves, and that the higher the energy of a particle, the smaller the wavelength of the corresponding wave. So the best answer we can give to our question depends on how high a particle energy we have at our disposal, because this determines on how small a length scale we can look. These particle energies are usually measured in units called electron volts. (In Thomson's experiments with electrons, we saw that he used an electric field to accelerate the electrons. The energy that an electron gains from an electric field of one volt is what is known as an electron volt.) In the nineteenth century, when the only particle energies that people knew how to use were the low energies of a few electron volts generated by chemical reactions such as burning, it was thought that atoms were the smallest unit. In Rutherford's experiment, the α-particles had energies of millions of electron volts. More recently, we have learned how to use electromagnetic fields to give particles energies of at first millions and then thousands of millions of electron volts. And so we know that particles that were thought to be "elementary" twenty years ago are, in fact, made up of smaller particles. May these, as we go to still higher energies, in turn be found to be made from still smaller particles? This is certainly possible, but we do have some theoretical reasons for believing that we have, or are very near to, a knowledge of the ultimate building blocks of nature.

Using the wave/particle duality discussed in the last chapter, everything in the universe, including light and gravity, can be described in terms of particles. These particles have a property called spin. One way of thinking of spin is to imagine the particles as little tops spinning about an axis. However, this can be misleading, because quantum mechanics tells us that the particles do not have any well-defined axis. What the spin of a particle really tells us is what the particle looks like from different directions. A particle of spin 0 is like a dot: it looks the same from every direction (Fig. 5.1–i). On the other hand, a particle of spin 1 is like an arrow: it looks different from different directions (Fig. 5.1–ii). Only if one turns it round a complete revolution (360 degrees) does the particle look the same. A particle of spin 2 is like a double-headed arrow (Fig. 5.1–iii): it looks the same if one turns it round half a revolution (180 degrees). Similarly, higher spin particles look the same if one turns them through smaller fractions of a complete revolution. All this seems fairly straightforward, but the remarkable fact is that there are particles that do not look the same if one turns them through

(i) SPIN = 0

(ii) SPIN = 1   (iii) SPIN = 2

FIGURE 5.1

just one revolution: you have to turn them through two complete revolutions! Such particles are said to have spin 1/2.

All the known particles in the universe can be divided into two groups: particles of spin 1/2, which make up the matter in the universe, and particles of spin 0, 1, and 2, which, we shall see, give rise to forces between the matter particles. The matter particles obey what is called Pauli's exclusion principle. This was discovered in 1925 by an Austrian physicist, Wolfgang Pauli—for which he received the Nobel prize in 1945. He was the archetypal theoretical physicist: it was said of him that even his presence in the same town would make experiments go wrong! Pauli's exclusion principle says that two similar particles cannot exist in the same state, that is, they cannot have both the same position and the same velocity, within the limits given by the uncertainty principle. The exclusion principle is crucial because it explains why matter particles do not collapse to a state of very high density under the influence of the forces produced by the particles of spin 0, 1, and 2: if the matter particles have very nearly the same positions, they must have different velocities, which means that they will not stay in the same position for long. If the world had been created without the exclusion principle, quarks would not form separate, well-defined protons and neutrons. Nor would these, together with electrons, form separate, well-defined atoms. They would all collapse to form a roughly uniform, dense "soup."

A proper understanding of the electron and other spin-1/2 particles did not come until 1928, when a theory was proposed by Paul Dirac, who later was elected to the Lucasian Professorship of Mathematics at Cambridge (the same professorship that Newton had once held and that I now hold). Dirac's theory was the first of its kind that was consistent with both quantum mechanics and the special theory of relativity. It explained mathematically why the electron had spin 1/2, that is, why it didn't look the same if you turned it through only one complete revolution, but did if you turned it through two revolutions. It also predicted that the electron should have a partner: an antielectron, or positron. The discovery of the positron in 1932 confirmed Dirac's theory and led to his being awarded the Nobel prize for physics in 1933. We now know that every particle has an antiparticle, with which it can annihilate. (In the case of the force-carrying particles, the antiparticles are the same as the particles themselves.) There could be whole antiworlds and antipeople made out of antiparticles. However, if you meet your antiself, don't shake hands! You would both vanish in a great flash of light. The question of why there seem to be so many more particles than antiparticles around us is extremely important, and I shall return to it later in the chapter.

In quantum mechanics, the forces or interactions between matter particles are all supposed to be carried by particles of integer spin—0, 1, or 2. What happens is that a matter particle, such as an electron or a quark, emits a force-carrying particle. The recoil from this emission changes the velocity of the matter particle. The force-carrying particle then collides with another matter particle and is absorbed. This collision changes the velocity of the second particle, just as if there had been a force between the two matter particles.

It is an important property of the force-carrying particles that they do not obey the exclusion principle. This means that there is no limit to the number that can be exchanged, and so they can give rise to a strong force. However, if the force-carrying particles have a high mass, it will be difficult to produce and exchange them over a large distance. So the forces that they carry will have only a short range. On the other hand, if the force-carrying particles have no mass of their own, the forces will be long range. The force-carrying particles exchanged between matter particles are said to be virtual particles because, unlike "real" particles, they cannot be directly detected by a particle detector. We know they exist, however, because they do have a measurable effect: they give rise to forces between matter particles. Particles of spin 0, 1, or 2 do also exist in some circumstances as real particles, when they can be directly detected. They then appear to us as what a classical physicist would call waves, such as waves of light or gravitational waves. They may sometimes be emitted when matter particles interact with each other by exchanging virtual force-carrying particles. (For example, the electric repulsive force between two electrons is due to the exchange of virtual photons, which can never be directly detected; but if one electron moves past another, real photons may be given off, which we detect as light waves.)

Force-carrying particles can be grouped into four categories according to the strength of the force that they carry and the particles with which they interact. It should be emphasized that this division into four classes is man-made; it is convenient for the construction of partial theories, but it may not correspond to anything deeper. Ultimately, most physicists hope to find a unified theory that will explain all four forces as different aspects of a single force. Indeed, many would say this is the prime goal of physics today. Recently, successful attempts have been made to unify three of the four categories of force—and I shall describe these in this chapter. The question of the unification of the remaining category, gravity, we shall leave till later.

The first category is the gravitational force. This force is universal, that is, every particle feels the force of gravity, according to its mass or energy. Gravity is the weakest of the four forces by a long way; it is so weak that we would not notice it at all were it not for two special properties that it has: it can act over large distances, and it is always attractive. This means that the very weak gravitational forces between the individual particles in two large bodies, such as the earth and the sun, can all add up to produce a significant force. The other three forces are either short range, or are sometimes attractive and sometimes repulsive, so they tend to cancel out. In the quantum mechanical way of looking at the gravitational field, the force between two matter particles is pictured as being carried by a particle of spin 2 called the graviton. This has no mass of its own, so the force that it carries is long range. The gravitational force between the sun and the earth is ascribed to the exchange of gravitons between the particles that make up these two bodies. Although the exchanged particles are virtual, they certainly do produce a measurable effect—they make the earth orbit the sun! Real gravitons make up what classical physicists would call gravitational waves, which are very weak—and so difficult to detect that they have never yet been observed.

The next category is the electromagnetic force, which interacts with electrically charged particles like electrons and quarks, but not with uncharged particles such as gravitons. It is much stronger than the gravitational force: the electromagnetic force between two electrons is about a million million million million million million million (1 with forty-two zeros after it) times bigger than the gravitational force. However, there are two kinds of electric charge, positive and negative. The force between two positive charges is repulsive, as is the force between two negative charges, but the force is attractive between a positive and a negative charge. A large body, such as the earth or the sun, contains nearly equal numbers of positive and negative charges. Thus the attractive and repulsive forces between the individual particles nearly cancel each other out, and there is very little net electromagnetic force. However, on the small scales of atoms and molecules, electromagnetic forces dominate. The electromagnetic attraction between negatively charged electrons and positively charged protons in the nucleus causes the electrons to orbit the nucleus of the atom, just as gravitational attraction causes the earth to orbit the sun. The electromagnetic attraction is pictured as being caused by the exchange of large numbers of virtual massless particles of spin 1, called photons. Again, the photons that are exchanged are virtual particles. However, when an electron changes from one allowed orbit to another one nearer to the nucleus, energy is released and a real photon is emitted—which can be observed as visible light by the human eye, if it has the right wavelength, or by a photon detector such as photographic film. Equally, if a real photon collides with an atom, it may move an electron from an orbit nearer the nucleus to one farther away. This uses up the energy of the photon, so it is absorbed.

The third category is called the weak nuclear force, which is responsible for radioactivity and which acts on all matter particles of spin 1/2, but not on particles of spin 0, 1, or 2, such as photons and gravitons. The weak nuclear force was not well understood until 1967, when Abdus Salam at Imperial College, London, and Steven Weinberg at Harvard both proposed theories that unified this interaction with the electromagnetic force, just as Maxwell had unified electricity and magnetism about a hundred years earlier. They suggested that in addition to the photon, there were three other spin-1 particles, known collectively as massive vector bosons, that carried the weak force. These were called W^+ (pronounced W plus), W^- (pronounced W minus), and Z^0 (pronounced Z naught), and each had a mass of around 100 GeV (GeV stands for giga-electron-volt, or one thousand million electron volts). The Weinberg–Salam theory exhibits a property known as spontaneous symmetry breaking. This means that what appear to be a number of completely different particles at low energies are in fact found to be all the same type of particle, only in different states. At high energies all these particles behave similarly. The effect is rather like the behavior of a roulette ball on a roulette wheel. At high energies (when the wheel is spun quickly) the ball behaves in essentially only one way—it rolls round and round. But as the wheel slows, the energy of the ball decreases, and eventually the ball drops into one of the thirty-seven slots in the wheel. In other words, at low energies there are thirty-seven different states in which the ball can exist. If, for some reason, we could only observe the ball at low energies, we would then think that there were thirty-seven different types of ball!

In the Weinberg–Salam theory, at energies much greater than 100 GeV, the three new particles and the photon would all behave in a similar manner. But at the lower particle energies that occur in most normal situations, this symmetry between the particles would be broken. W^+, W^-, and Z^0

would acquire large masses, making the forces they carry have a very short range. At the time that Salam and Weinberg proposed their theory, few people believed them, and particle accelerators were not powerful enough to reach the energies of 100 GeV required to produce real W^+, W^-, or Z^0 particles. However, over the next ten years or so, the other predictions of the theory at lower energies agreed so well with experiment that, in 1979, Salam and Weinberg were awarded the Nobel prize for physics, together with Sheldon Glashow, also at Harvard, who had suggested similar unified theories of the electromagnetic and weak nuclear forces. The Nobel committee was spared the embarrassment of having made a mistake by the discovery in 1983 at CERN (European Centre for Nuclear Research) of the three massive partners of the photon, with the correct predicted masses and other properties. Carlo Rubbia, who led the team of several hundred physicists that made the discovery, received the Nobel prize in 1984, along with Simon van der Meer, the CERN engineer who developed the antimatter storage system employed. (It is very difficult to make a mark in experimental physics these days unless you are already at the top!)

The fourth category is the strong nuclear force, which holds the quarks together in the proton and neutron, and holds the protons and neutrons together in the nucleus of an atom. It is believed that this force is carried by another spin-1 particle, called the gluon, which interacts only with itself and with the quarks. The strong nuclear force has a curious property called confinement: it always binds particles together into combinations that have no color. One cannot have a single quark on its own because it would have a color (red, green, or blue). Instead, a red quark has to be joined to a green and a blue quark by a "string" of gluons (red + green + blue = white). Such a triplet constitutes a proton or a neutron. Another possibility is a pair consisting of a quark and an antiquark (red + antired, or green + antigreen, or blue + antiblue = white). Such combinations make up the particles known as mesons, which are unstable because the

FIGURE 5.2
A proton and an antiproton collide at high energy, producing a couple of almost free quarks.

quark and antiquark can annihilate each other, producing electrons and other particles. Similarly, confinement prevents one having a single gluon on its own, because gluons also have color. Instead, one has to have a collection of gluons whose colors add up to white. Such a collection forms an unstable particle called a glueball.

The fact that confinement prevents one from observing an isolated quark or gluon might seem to make the whole notion of quarks and gluons as particles somewhat metaphysical. However, there is another property of the strong nuclear force, called asymptotic freedom, that makes the concept of quarks and gluons well-defined. At normal energies, the strong nuclear force is indeed strong, and it binds the quarks tightly together. However, experiments with large particle accelerators indicate that at high energies the strong force becomes much weaker, and the quarks and gluons behave almost like free particles. Fig. 5.2 shows a photograph of a collision between a high energy proton and antiproton. Several almost free quarks were produced and gave rise to the "jets" of tracks seen in the picture.

The success of the unification of the electromagnetic and weak nuclear forces led to a number of attempts to combine these two forces with the strong nuclear force into what is called a grand unified theory (or GUT). This title is rather an exaggeration: the resultant theories are not all that grand, nor are they fully unified, as they do not include gravity. Nor are they really complete theories, because they contain a number of parameters whose values cannot be predicted from the theory but have to be chosen to fit in with experiment. Nevertheless, they may be a step toward a complete, fully unified theory. The basic idea of GUTs is as follows: As was mentioned above, the strong nuclear force gets weaker at high energies. On the other hand, the electromagnetic and weak forces, which are not asymptotically free, get stronger at high energies. At some very high energy, called the grand unification energy, these three forces would all have the same strength and so could just be different aspects of a single force. The GUTs also predict that at this energy the different spin-1/2 matter particles, like quarks and electrons, would also all be essentially the same, thus achieving another unification.

The value of the grand unification energy is not very well known, but it would probably have to be at least a thousand million million GeV. The present generation of particle accelerators can collide particles at energies of about one hundred GeV, and machines are planned that would raise this to a few thousand GeV. But a machine that was powerful enough to accelerate particles to the grand unification energy would have to be as big as the Solar System—and would be unlikely to be funded in the present economic climate. Thus it is impossible to test grand unified theories directly in the laboratory. However, just as in the case of the electromagnetic and weak unified theory, there are low-energy consequences of the theory that can be tested.

The most interesting of these is the prediction that protons, which make up much of the mass of ordinary matter, can spontaneously decay into lighter particles such as antielectrons. The reason this is possible is that at the grand unification energy there is no essential difference between a quark and an antielectron. The three quarks inside a proton normally do not have enough energy to change into antielectrons, but very occasionally one of them may acquire sufficient energy to make the transition because the uncertainty principle means that the energy of the quarks inside the proton cannot be fixed exactly. The proton would then decay. The probability of a quark gaining sufficient energy is so low that one is likely to have to wait at least a million million million million million years (1 followed by thirty zeros). This is much longer than the time since the big bang, which is a mere ten thousand million years or so (1 followed by ten zeros). Thus one might think that the possibility of spontaneous proton decay could not be tested experimentally. However, one can increase one's chances of detecting a decay by observing a large amount of matter containing a very large number of protons. (If, for example, one observed a number of protons equal to 1 followed by thirty-one zeros for a period of one year, one would expect, according to the simplest GUT, to observe more than one proton decay.)

A number of such experiments have been carried out, but none have yielded definite evidence of proton or neutron decay. One experiment used eight thousand tons of water and was performed in the Morton Salt Mine in Ohio (to avoid other events taking place, caused by cosmic rays, that might be confused with proton decay). Since no spontaneous proton decay had been observed during the experiment, one can calculate that the probable life of the proton must be greater than ten million million million million million years (1 with thirty-one zeros). This is longer than the lifetime predicted by the simplest grand unified theory, but there are more elaborate theories in which the predicted lifetimes are longer. Still more sensitive experiments involving even larger quantities of matter will be needed to test them.

Even though it is very difficult to observe spontaneous proton decay, it may be that our very existence is a consequence of the reverse process, the production of protons, or more simply, of quarks, from an initial situation in which there were no more quarks than antiquarks, which is the most natural way to imagine the universe starting out. Matter on the earth is made up mainly of protons and neutrons, which in turn are made up of quarks. There are no antiprotons or antineutrons, made up from antiquarks, except for a few that physicists produce in large particle accelerators. We have evidence from cosmic rays that the same is true for all the matter in our galaxy: there are no antiprotons or antineutrons apart from a small number that are produced as particle/antiparticle pairs in high-energy collisions. If there were large regions of antimatter in our galaxy, we would expect to observe large quantities of radiation from the borders between the regions of matter and antimatter, where many particles would be colliding with their antiparticles, annihilating each other and giving off high energy radiation.

We have no direct evidence as to whether the matter in other galaxies is made up of protons and neutrons or antiprotons and antineutrons, but it must be one or the other: there cannot be a mixture in a single galaxy because in that case we would again observe a lot of radiation from annihilations. We therefore believe that all galaxies are composed of quarks rather than antiquarks; it seems implausible that some galaxies should be matter and some antimatter.

Why should there be so many more quarks than antiquarks? Why are there not equal numbers of each? It is certainly fortunate for us that the numbers are unequal because, if they had been the same, nearly all the quarks and antiquarks would have annihilated each other in the early universe and left a universe filled with radiation but hardly any matter. There would then have been no galaxies, stars, or planets on which human life could have developed. Luckily, grand unified theories may provide an explanation of why the universe should now contain more quarks than antiquarks, even if it started out with equal numbers of each. As we have seen, GUTs allow quarks to change into antielectrons at high energy. They also allow the reverse processes, antiquarks turning into electrons, and electrons and antielectrons turning into antiquarks and quarks. There was a time in the very early universe when it was so hot that the particle energies would have been high enough for these transformations to take place. But why should that lead to more quarks than anti-quarks? The reason is that the laws of physics are not quite the same for particles and antiparticles.

Up to 1956 it was believed that the laws of physics obeyed each of three separate symmetries called C, P, and T. The symmetry C means that the laws are the same for particles and antiparticles. The symmetry P means that the laws are the same for any situation and its mirror image (the mirror image of a particle spinning in a right-handed direction is one spinning in a left-handed direction). The symmetry T means that if you reverse the direction of motion of all particles and antiparticles, the system should go back to what it was at earlier times; in other words, the laws are the same in the forward and backward directions of time.

In 1956 two American physicists, Tsung-Dao Lee and Chen Ning Yang, suggested that the weak force does not in fact obey the symmetry P. In other words, the weak force would make the universe develop in a different way from the way in which the mirror image of the universe would develop. The same year, a colleague, Chien-Shiung Wu, proved their prediction correct. She did this by lining up the nuclei of radioactive atoms in a magnetic field, so that they were all spinning in the same direction, and showed that the electrons were given off more in one direction than another. The following year, Lee and Yang received the Nobel prize for their idea. It was also found that the weak force did not obey the symmetry C. That is, it would cause a universe composed of antiparticles to behave differently from our universe. Nevertheless, it seemed that the weak force did obey the combined symmetry CP. That is, the universe would develop in the same way as its mirror image if, in addition, every particle was swapped with its antiparticle! However, in 1964 two more Americans, J. W. Cronin and Val Fitch, discovered that even the CP symmetry was not obeyed in the decay of certain particles called K-mesons. Cronin and Fitch eventually received the Nobel prize for their work in 1980. (A lot of prizes have been awarded for showing that the universe is not as simple as we might have thought!)

There is a mathematical theorem that says that any theory that obeys quantum mechanics and relativity must always obey the combined symmetry CPT. In other words, the universe would have to behave the same if one replaced particles by antiparticles, took the mirror image, and also reversed the direction of time. But Cronin and Fitch showed that if one replaces particles by antiparticles and takes the mirror image, but does not reverse the direction of time, then the universe does *not* behave the same. The laws of physics, therefore, must change if one reverses the direction of time—they do not obey the symmetry T.

Certainly the early universe does not obey the symmetry T: as time runs forward the universe expands—if it ran backward, the universe would be contracting. And since there are forces that do not obey the symmetry T, it follows that as the universe expands, these forces could cause more antielectrons to turn into quarks than electrons into antiquarks. Then, as the universe expanded and cooled, the antiquarks would annihilate with the quarks, but since there would be more quarks than antiquarks, a small excess of quarks would remain. It is these that make up the matter we see today and out of which we ourselves are made. Thus our very existence could be regarded as a confirmation of grand unified theories, though a qualitative one only; the uncertainties are such that one cannot predict the numbers of quarks that will be left after the annihilation, or even whether it would be quarks or antiquarks that would remain. (Had it been an excess of antiquarks, however, we would simply have named antiquarks quarks, and quarks antiquarks.)

Grand unified theories do not include the force of gravity. This does not matter too much, because gravity is such a weak force that its effects can usually be neglected when we are dealing with elementary particles or atoms. However, the fact that it is both long range and always attractive means that its effects all add up. So for a sufficiently large number of matter particles, gravitational forces can dominate over all other forces. This is why it is gravity that determines the evolution of the universe. Even for objects the size of stars, the attractive force of gravity can win over all the other forces and cause the star to collapse. My work in the 1970s focused on the black holes that can result from such stellar collapse and the intense gravitational fields around them. It was this that led to the first hints of how the theories of quantum mechanics and general relativity might affect each other—a glimpse of the shape of a quantum theory of gravity yet to come.

6

Black Holes

The term *black hole* is of very recent origin. It was coined in 1969 by the American scientist John Wheeler as a graphic description of an idea that goes back at least two hundred years, to a time when there were two theories about light: one, which Newton favored, was that it was composed of particles; the other was that it was made of waves. We now know that really both theories are correct. By the wave/particle duality of quantum mechanics, light can be regarded as both a wave and a particle. Under the theory that light is made up of waves, it was not clear how it would respond to gravity. But if light is composed of particles, one might expect them to be affected by gravity in the same way that cannonballs, rockets, and planets are. At first people thought that particles of light traveled infinitely fast, so gravity would not have been able to slow them down, but the discovery by Roemer that light travels at a finite speed meant that gravity might have an important effect.

On this assumption, a Cambridge don, John Michell, wrote

a paper in 1783 in the *Philosophical Transactions of the Royal Society of London* in which he pointed out that a star that was sufficiently massive and compact would have such a strong gravitational field that light could not escape: any light emitted from the surface of the star would be dragged back by the star's gravitational attraction before it could get very far. Michell suggested that there might be a large number of stars like this. Although we would not be able to see them because the light from them would not reach us, we would still feel their gravitational attraction. Such objects are what we now call black holes, because that is what they are: black voids in space. A similar suggestion was made a few years later by the French scientist the Marquis de Laplace, apparently independently of Michell. Interestingly enough, Laplace included it in only the first and second editions of his book *The System of the World,* and left it out of later editions; perhaps he decided that it was a crazy idea. (Also, the particle theory of light went out of favor during the nineteenth century; it seemed that everything could be explained by the wave theory, and according to the wave theory, it was not clear that light would be affected by gravity at all.)

In fact, it is not really consistent to treat light like cannonballs in Newton's theory of gravity because the speed of light is fixed. (A cannonball fired upward from the earth will be slowed down by gravity and will eventually stop and fall back; a photon, however, must continue upward at a constant speed. How then can Newtonian gravity affect light?) A consistent theory of how gravity affects light did not come along until Einstein proposed general relativity in 1915. And even then it was a long time before the implications of the theory for massive stars were understood.

To understand how a black hole might be formed, we first need an understanding of the life cycle of a star. A star is formed when a large amount of gas (mostly hydrogen) starts to collapse in on itself due to its gravitational attraction. As it contracts the atoms of the gas collide with each other more and more frequently and at greater and greater speeds—the gas heats up. Eventually, the gas will be so hot that when the hydrogen atoms collide they no longer bounce off each other, but instead coalesce to form helium. The heat released in this reaction, which is like a controlled hydrogen bomb explosion, is what makes the star shine. This additional heat also increases the pressure of the gas until it is sufficient to balance the gravitational attraction, and the gas stops contracting. It is a bit like a balloon—there is a balance between the pressure of the air inside, which is trying to make the balloon expand, and the tension in the rubber, which is trying to make the balloon smaller. Stars will remain stable like this for a long time, with heat from the nuclear reactions balancing the gravitational attraction. Eventually, however, the star will run out of its hydrogen and other nuclear fuels. Paradoxically, the more fuel a star starts off with, the sooner it runs out. This is because the more massive the star is, the hotter it needs to be to balance its gravitational attraction. And the hotter it is, the faster it will use up its fuel. Our sun has probably got enough fuel for another five thousand million years or so, but more massive stars can use up their fuel in as little as one hundred million years, much less than the age of the universe. When a star runs out of fuel, it starts to cool off and so to contract. What might happen to it then was first understood only at the end of the 1920s.

In 1928 an Indian graduate student, Subrahmanyan Chandrasekhar, set sail for England to study at Cambridge with the British astronomer Sir Arthur Eddington, an expert on general relativity. (According to some accounts, a journalist told Eddington in the early 1920s that he had heard there were only three people in the world who understood general relativity. Eddington paused, then replied, "I am trying to think who the third person is.") During his voyage from India, Chandrasekhar worked out how big a star could be and still support itself against its own gravity after it had used up all its fuel. The idea was this: When the star becomes small, the matter particles get very near each other, and so according to the Pauli exclusion principle, they must have very different velocities. This makes them move away from each other and so tends to make the star expand. A star can therefore maintain itself at a constant radius by a balance between the attraction of gravity and the repulsion that arises from the exclusion principle, just as earlier in its life gravity was balanced by the heat.

Chandrasekhar realized, however, that there is a limit to the repulsion that the exclusion principle can provide. The theory of relativity limits the maximum difference in the velocities of the matter particles in the star to the speed of light. This means that when the star got sufficiently dense, the repulsion caused by the exclusion principle would be less than the attraction of gravity. Chandrasekhar calculated that a cold star of more than about one and a half times the mass of the sun would not be able to support itself against its own gravity. (This mass is now known as the Chandrasekhar limit.) A similar discovery was made about the same time by the Russian scientist Lev Davidovich Landau.

This had serious implications for the ultimate fate of massive stars. If a star's mass is less than the Chandrasekhar limit, it can eventually stop contracting and settle down to a possible final state as a "white dwarf" with a radius of a few thousand miles and a density of hundreds of tons per cubic inch. A white dwarf is supported by the exclusion principle repulsion between the electrons in its matter. We observe a large number of these white dwarf stars. One of the first to be discovered is a star that is orbiting around Sirius, the brightest star in the night sky.

Landau pointed out that there was another possible final state for a star, also with a limiting mass of about one or two times the mass of the sun but much smaller even than a white dwarf. These stars would be supported by the exclusion principle repulsion between neutrons and protons, rather than between electrons. They were therefore called neutron stars. They would have a radius of only ten miles or so and a density of hundreds of millions of tons per cubic inch. At the time they were first predicted, there was no way that neutron stars could be observed. They were not actually detected until much later.

Stars with masses above the Chandrasekhar limit, on the other hand, have a big problem when they come to the end of their fuel. In some cases they may explode or manage to throw off enough matter to reduce their mass below the limit and so avoid catastrophic gravitational collapse, but it was difficult to believe that this always happened, no matter how big the star. How would it know that it had to lose weight? And even if every star managed to lose enough mass to avoid collapse, what would happen if you added more mass to a white dwarf or neutron star to take it over the limit? Would it collapse to infinite density? Eddington was shocked by that implication, and he refused to believe Chandrasekhar's result. Eddington thought it was simply not possible that a star could collapse to a point. This was the view of most scientists: Einstein himself wrote a paper in which he claimed that stars would not shrink to zero size. The hostility of other scientists, particularly Eddington, his former teacher and the leading authority on the structure of stars, persuaded Chandrasekhar to abandon this line of work and turn instead to other problems in astronomy, such as the motion of star clusters. However, when he was awarded the Nobel prize in 1983, it was, at least in part, for his early work on the limiting mass of cold stars.

Chandrasekhar had shown that the exclusion principle could not halt the collapse of a star more massive than the Chandrasekhar limit, but the problem of understanding what would happen to such a star, according to general relativity, was first solved by a young American, Robert Oppenheimer, in 1939. His result, however, suggested that there would be no observational consequences that could be detected by the telescopes of the day. Then World War II intervened and Oppenheimer himself became closely involved in the atom bomb project. After the war the problem of gravitational collapse was largely forgotten as most scientists became caught up in what happens on the scale of the atom and its nucleus.

In the 1960s, however, interest in the large-scale problems of astronomy and cosmology was revived by a great increase in the number and range of astronomical observations brought about by the application of modern technology. Oppenheimer's work was then rediscovered and extended by a number of people.

The picture that we now have from Oppenheimer's work is as follows: The gravitational field of the star changes the paths of light rays in space-time from what they would have been had the star not been present. The light cones, which indicate the paths followed in space and time by flashes of light emitted from their tips, are bent slightly inward near the surface of the star. This can be seen in the bending of light from distant stars observed during an eclipse of the sun. As the star contracts, the gravitational field at its surface gets stronger and the light cones get bent inward more. This makes it more difficult for light from the star to escape, and the light appears dimmer and redder to an observer at a distance. Eventually, when the star has shrunk to a certain critical radius, the gravitational field at the surface becomes so strong that the light cones are bent inward so much that light can no longer escape (Fig. 6.1). According to the theory of relativity, nothing can travel faster than light. Thus if light cannot escape, neither can anything else; everything is dragged back by the gravitational field. So one has a set of events, a region of space-time, from which it is not possible to escape to reach a distant observer. This region is what we now call a black hole. Its boundary is called the event horizon and it coincides with the paths of light rays that just fail to escape from the black hole.

In order to understand what you would see if you were watching a star collapse to form a black hole, one has to remember that in the theory of relativity there is no absolute time. Each observer has his own measure of time. The time for someone on a star will be different from that for someone at a distance, because of the gravitational field of the star. Suppose an intrepid astronaut on the surface of the collapsing star, collapsing inward with it, sent a signal every second, according to his watch, to his spaceship orbiting about the star. At some time on his watch, say 11:00, the star would shrink below the critical radius at which the gravitational field becomes so strong nothing can escape, and his signals would no longer reach the spaceship. As 11:00 approached, his companions watching from the spaceship would find the intervals between successive signals from the astronaut getting longer and longer, but this effect would be very small before 10:59:59. They would have to wait only very slightly more than a second between the astronaut's 10:59:58 signal and the one that he sent when his watch read 10:59:59, but they would have to wait forever for the 11:00 signal. The light waves emitted from the surface of the star between 10:59:59 and 11:00, by the astronaut's watch, would be spread out over an infinite period of time, as seen from the spaceship. The time interval between the arrival of successive waves at the spaceship would get longer and longer, so the light from the star would appear redder and redder and fainter and fainter. Eventually, the star would be so dim that it could no longer be seen from the spaceship: all that would be left would be a black hole in space. The star would, however, continue to exert the same gravitational force on the spaceship, which would continue to orbit the black hole.

This scenario is not entirely realistic, however, because of the following problem. Gravity gets weaker the farther you are from the star, so the gravitational force on our intrepid astronaut's feet would always be greater than the force on his head. This difference in the forces would stretch our astronaut out like spaghetti or tear him apart before the star had contracted to the critical radius at which the event horizon formed! However, we believe that there are much larger objects in the universe, like the central regions of galaxies, that can also undergo gravitational collapse to produce black holes; an astronaut on one of these would not be torn apart before the black hole formed. He would not, in fact, feel anything special as he reached the critical radius, and could pass the

SINGULARITY

LIGHT RAY EMITTED
AT THE TIME
THE EVENT HORIZON FORMS

LIGHT RAY EMITTED
AFTER EVENT HORIZON FORMS

LIGHT CONES

LIGHT RAY EMITTED
BEFORE EVENT HORIZON FORMS

INTERIOR OF STAR

TIME

0

DISTANCE FROM CENTER OF STAR

FIGURE 6.1

FIGURE 6.1

point of no return without noticing it. However, within just a few hours, as the region continued to collapse, the difference in the gravitational forces on his head and his feet would become so strong that again it would tear him apart.

The work that Roger Penrose and I did between 1965 and 1970 showed that, according to general relativity, there must be a singularity of infinite density and space-time curvature within a black hole. This is rather like the big bang at the beginning of time, only it would be an end of time for the collapsing body and the astronaut. At this singularity the laws of science and our ability to predict the future would break down. However, any observer who remained outside the black hole would not be affected by this failure of predictability, because neither light nor any other signal could reach him from the singularity. This remarkable fact led Roger Penrose to propose the cosmic censorship hypothesis, which might be paraphrased as "God abhors a naked singularity." In other words, the singularities produced by gravitational collapse occur only in places, like black holes, where they are decently hidden from outside view by an event horizon. Strictly, this is what is known as the weak cosmic censorship hypothesis: it protects observers who remain outside the black hole from the consequences of the breakdown of predictability that occurs at the singularity, but it does nothing at all for the poor unfortunate astronaut who falls into the hole.

There are some solutions of the equations of general relativity in which it is possible for our astronaut to see a naked singularity: he may be able to avoid hitting the singularity and instead fall through a "wormhole" and come out in another region of the universe. This would offer great possibilities for travel in space and time, but unfortunately it seems that these solutions may all be highly unstable; the least disturbance, such as the presence of an astronaut, may change them so that the astronaut could not see the singularity until he hit it and his time came to an end. In other words, the singularity would always lie in his future and never in his past. The strong version of the cosmic censorship hypothesis states that in a realistic solution, the singularities would always lie either entirely in the future (like the singularities of gravitational collapse) or entirely in the past (like the big bang). It is greatly to be hoped that some version of the censorship hypothesis holds because close to naked singularities it may be possible to travel into the past. While this would be fine for writers of science fiction, it would mean that no one's life would ever be safe: someone might go into the past and kill your father or mother before you were conceived!

The event horizon, the boundary of the region of space-time from which it is not possible to escape, acts rather like a one-way membrane around the black hole: objects, such as unwary astronauts, can fall through the event horizon into the black hole, but nothing can ever get out of the black hole through the event horizon. (Remember that the event horizon is the path in space-time of light that is trying to escape from the black hole, and nothing can travel faster than light.) One could well say of the event horizon what the poet Dante said of the entrance to Hell: "All hope abandon, ye who enter here." Anything or anyone who falls through the event horizon will soon reach the region of infinite density and the end of time.

General relativity predicts that heavy objects that are moving will cause the emission of gravitational waves, ripples in the curvature of space that travel at the speed of light. These are similar to light waves, which are ripples of the electromagnetic field, but they are much harder to detect. Like light, they carry energy away from the objects that emit them. One would therefore expect a system of massive objects to settle down eventually to a stationary state, because the energy in any movement would be carried away by the emission of gravitational waves. (It is rather like dropping a cork into water: at first it bobs up and down a great deal, but as the ripples carry away its energy, it eventually settles down to a stationary state.) For example, the movement of the earth in its orbit round the sun produces gravitational waves. The effect of the energy loss will be to change the orbit of the earth so that gradually it gets nearer and nearer to the sun, eventually collides with it, and settles down to a stationary state. The rate of energy loss in the case of the earth and the sun is very low—about enough to run a small electric heater. This means it will take about a thousand million million million million years for the earth to run into the sun, so there's no immediate cause for worry! The change in the orbit of the earth is too slow to be observed, but this same effect has been observed over the past few years occurring in the system called PSR 1913 + 16 (*PSR* stands for "pulsar," a special type of neutron star that emits regular pulses of radio waves). This system contains two neutron stars orbiting each other, and the energy they are losing by the emission of gravitational waves is causing them to spiral in toward each other.

During the gravitational collapse of a star to form a black hole, the movements would be much more rapid, so the rate at which energy is carried away would be much higher. It would therefore not be too long before it settled down to a stationary state. What would this final stage look like? One might suppose that it would depend on all the complex features of the star from which it had formed—not only its mass and rate of rotation, but also the different densities of various parts of the star, and the complicated movements of the gases within the star. And if black holes were as varied as the objects that collapsed to form them, it might be very difficult to make any predictions about black holes in general.

In 1967, however, the study of black holes was revolutionized by Werner Israel, a Canadian scientist (who was born in Berlin, brought up in South Africa, and took his doctoral degree in Ireland). Israel showed that, according to general relativity, nonrotating black holes must be very simple; they were perfectly spherical, their size depended only on their mass, and any two such black holes with the same mass were identical. They could, in fact, be described by a particular solution of Einstein's equations that had been known since 1917, found by Karl Schwarzschild shortly after the discovery of general relativity. At first many people, including Israel himself, argued that since black holes had to be perfectly spherical, a black hole could only form from the collapse of a perfectly spherical object. Any real star—which would never be *perfectly* spherical—could therefore only collapse to form a naked singularity.

There was, however, a different interpretation of Israel's result, which was advocated by Roger Penrose and John Wheeler in particular. They argued that the rapid movements involved in a star's collapse would mean that the gravitational waves it gave off would make it ever more spherical, and by the time it had settled down to a stationary state, it would be precisely spherical. According to this view, any nonrotating star, however complicated its shape and internal structure, would end up after gravitational collapse as a perfectly spherical black hole, whose size would depend only on its mass. Further calculations supported this view, and it soon came to be adopted generally.

Israel's result dealt with the case of black holes formed from nonrotating bodies only. In 1963, Roy Kerr, a New Zealander, found a set of solutions of the equations of general relativity that described rotating black holes. These "Kerr" black holes rotate at a constant rate, their size and shape depending only on their mass and rate of rotation. If the rotation is zero, the black hole is perfectly round and the solution is identical to the Schwarzschild solution. If the rotation is nonzero, the black hole bulges outward near its equator (just as the earth or the sun bulge due to their rotation), and the faster it rotates, the more it bulges. So, to extend Israel's result to include rotating bodies, it was conjectured that any rotating body that collapsed to form a black hole would eventually settle down to a stationary state described by the Kerr solution.

In 1970 a colleague and fellow research student of mine at Cambridge, Brandon Carter, took the first step toward proving this conjecture. He showed that, provided a stationary rotating black hole had an axis of symmetry, like a spinning top, its size and shape would depend only on its mass and rate of rotation. Then, in 1971, I proved that any stationary rotating black hole would indeed have such an axis of symmetry. Finally, in 1973, David Robinson at Kings College, London, used Carter's and my results to show that the conjecture had been correct: such a black hole had indeed to be the Kerr solution. So after gravitational collapse a black hole must settle down into a state in which it could be rotating, but not pulsating. Moreover, its size and shape would depend only on its mass and rate of rotation, and not on the nature of the body that had collapsed to form it. This result became known by the maxim: "A black hole has no hair." The "no hair" theorem is of great practical importance, because it so greatly restricts the possible types of black holes. One can therefore make detailed models of objects that might contain black holes and compare the predictions of the models with observations. It also means that a very large amount of information about the body that has collapsed must be lost when a black hole is formed, because afterward all we can possibly measure about the body is its mass and rate of rotation. The significance of this will be seen in the next chapter.

Black holes are one of only a fairly small number of cases in the history of science in which a theory was developed in great detail as a mathematical model before there was any evidence from observations that it was correct. Indeed, this used to be the main argument of opponents of black holes: how could one believe in objects for which the only evidence was calculations based on the dubious theory of general relativity? In 1963, however, Maarten Schmidt, an astronomer at the Palomar Observatory in California, measured the red shift of a faint starlike object in the direction of the source of radio waves called 3C273 (that is, source number 273 in the third Cambridge catalogue of radio sources). He found it was too large to be caused by a gravitational field: if it had been a gravitational red shift, the object would have to be so massive and so near to us that it would disturb the orbits of planets in the Solar System. This suggested that the red shift was instead caused by the expansion of the universe, which, in turn, meant that the object was a very long distance away. And to be visible at such a great distance, the object must be very bright, must, in other words, be emitting a huge amount of energy. The only mechanism that people could think of that would produce such large quantities of energy seemed to be the gravitational collapse not just of a star but of a whole central region of a galaxy. A number of other, similar "quasi-stellar objects," or quasars, have been discovered, all with large red shifts. But they are all too far away and therefore too difficult to observe to provide conclusive evidence of black holes.

Further encouragement for the existence of black holes came in 1967 with the discovery by a research student at Cambridge, Jocelyn Bell, of objects in the sky that were emitting regular pulses of radio waves. At first Bell and her supervisor, Antony Hewish, thought they might have made contact with an alien civilization in the galaxy! Indeed, at the seminar at which they announced their discovery, I remember that they called the first four sources to be found LGM 1–4, *LGM* standing for "Little Green Men." In the end, however, they and everyone else came to the less romantic conclusion that these objects, which were given the name *pulsars,* were in fact rotating neutron stars that were emitting pulses of radio waves because of a complicated interaction between their magnetic fields and surrounding matter. This was bad news for writers of space westerns, but very hopeful for the small number of us who believed in black holes at that time: it was the first positive evidence that neutron stars existed. A neutron star has a radius of about ten miles, only a few times the critical radius at which a star becomes a black hole. If a star could collapse to such a small size, it is not unreasonable to expect that other stars could collapse to even smaller size and become black holes.

How could we hope to detect a black hole, as by its very definition it does not emit any light? It might seem a bit like looking for a black cat in a coal cellar. Fortunately, there is a way. As John Michell pointed out in his pioneering paper in 1783, a black hole still exerts a gravitational force on nearby objects. Astronomers have observed many systems in which two stars orbit around each other, attracted toward each other by gravity. They also observe systems in which there is only one visible star that is orbiting around some unseen companion. One cannot, of course, immediately conclude that the companion is a black hole: it might merely be a star that is too faint to be seen. However, some of these systems, like the one called Cygnus X-1 (Fig. 6.2), are also strong sources of X rays. The best explanation for this phenomenon is that matter has been blown off the surface of the visible star. As it falls toward the unseen companion, it develops a spiral motion (rather like water running out of a bath), and it gets very hot, emitting X rays (Fig. 6.3). For this mechanism to work, the unseen object has to be very small, like a white dwarf, neutron star, or black hole. From the observed orbit of the visible star, one can determine the lowest possible mass of the unseen object. In the case of Cygnus X-1, this is about six times the mass of the sun, which, according to Chandrasekhar's result, is too great for the unseen object to be a white dwarf. It is also too large a mass to be a neutron star. It seems, therefore, that it must be a black hole.

There are other models to explain Cygnus X-1 that do not include a black hole, but they are all rather farfetched. A black hole seems to be the only really natural explanation of the observations. Despite this, I have a bet with Kip Thorne of the California Institute of Technology that in fact Cygnus X-1 does not contain a black hole! This is a form of insurance policy for me. I have done a lot of work on black holes, and it would all be wasted if it turned out that black holes do not exist. But in that case, I would have the consolation of winning my bet, which would bring me four years of the magazine *Private Eye*. If black holes do exist, Kip will get one year of *Penthouse*. When we made the bet, in 1975, we were 80 percent certain that Cygnus was a black hole. By now, I would say that we are about 95 percent certain, but the bet has yet to be settled.

We also now have evidence for several other black holes in systems like Cygnus X-1 in our galaxy and in two neighboring galaxies called the Magellanic Clouds. The number of black holes, however, is almost certainly very much higher; in the long history of the universe, many stars must have burned all their nuclear fuel and have had to collapse. The number of black holes may well be greater even than the number of visible stars, which totals about a hundred thousand million in our galaxy alone. The extra gravitational attraction of such a large number of black holes could explain why our galaxy rotates at the rate it does: the mass of the visible stars is insufficient to account for this. We also have some evidence that there is a much larger black hole, with a mass of about a hundred thousand times that of the sun, at the center of our galaxy. Stars in the galaxy that come too near this black hole will be torn apart by the difference in the gravitational forces on their near and far sides. Their remains, and gas that is thrown off other stars, will fall toward the black hole. As in the case of Cygnus X-1, the gas will spiral inward and will heat up, though not as much as in that case. It will not get hot enough to emit X rays, but it could account for the very compact source of radio waves and infrared rays that is observed at the galactic center.

It is thought that similar but even larger black holes, with masses of about a hundred million times the mass of the sun, occur at the centers of quasars. Matter falling into such a supermassive black hole would provide the only source of power great enough to explain the enormous amounts of energy that these objects are emitting. As the matter spirals into the black hole, it would make the black hole rotate in the same direction, causing it to develop a magnetic field rather like that of the earth. Very high energy particles would be generated near the black hole by the in-falling matter. The magnetic field would be so strong that it could focus these particles into jets ejected outward along the axis of rotation of

FIGURE 6.2

The brighter of the two stars near the center of the photograph is Cygnus X-1, which is thought to consist of a black hole and a normal star, orbiting around each other.

the black hole, that is, in the directions of its north and south poles. Such jets are indeed observed in a number of galaxies and quasars.

One can also consider the possibility that there might be black holes with masses much less than that of the sun. Such black holes could not be formed by gravitational collapse, because their masses are below the Chandrasekhar mass limit: stars of this low mass can support themselves against the force of gravity even when they have exhausted their nuclear fuel. Low-mass black holes could form only if matter was compressed to enormous densities by very large external pressures. Such conditions could occur in a very big hydrogen bomb: the physicist John Wheeler once calculated that if

FIGURE 6.3

one took all the heavy water in all the oceans of the world, one could build a hydrogen bomb that would compress matter at the center so much that a black hole would be created. (Of course, there would be no one left to observe it!) A more practical possibility is that such low-mass black holes might have been formed in the high temperatures and pressures of the very early universe. Black holes would have been formed only if the early universe had not been perfectly smooth and uniform, because only a small region that was denser than average could be compressed in this way to form a black hole. But we know that there must have been some irregularities, because otherwise the matter in the universe would still be perfectly uniformly distributed at the present epoch, instead of being clumped together in stars and galaxies.

Whether the irregularities required to account for stars and galaxies would have led to the formation of a significant number of "primordial" black holes clearly depends on the details of the conditions in the early universe. So if we could determine how many primordial black holes there are now, we would learn a lot about the very early stages of the universe. Primordial black holes with masses more than a thousand million tons (the mass of a large mountain) could be detected only by their gravitational influence on other, visible matter or on the expansion of the universe. However, as we shall learn in the next chapter, black holes are not really black after all: they glow like a hot body, and the smaller they are, the more they glow. So, paradoxically, smaller black holes might actually turn out to be easier to detect than large ones!

7

Black Holes
Ain't So Black

Before 1970, my research on general relativity had concentrated mainly on the question of whether or not there had been a big bang singularity. However, one evening in November that year, shortly after the birth of my daughter, Lucy, I started to think about black holes as I was getting into bed. My disability makes this rather a slow process, so I had plenty of time. At that date there was no precise definition of which points in space-time lay inside a black hole and which lay outside. I had already discussed with Roger Penrose the idea of defining a black hole as the set of events from which it was not possible to escape to a large distance, which is now the generally accepted definition. It means that the boundary of the black hole, the event horizon, is formed by the paths in space-time of rays of light that just fail to get away from the black hole, hovering forever just on the edge (Fig. 7.1). It is a bit like running away from the police and just managing to keep one step ahead but not being able to get clear away!

Suddenly I realized that the paths of these light rays could

TIME

SINGULARITY

LIGHT RAY
HITS SINGULARITY

EVENT
HORIZON

LIGHT RAY ESCAPES

INTERIOR
OF
BLACK
HOLE

LIGHT RAY
AT EVENT HORIZON

EVENT
HORIZON

0

DISTANCE FROM SINGULARITY

FIGURE 7.1

never approach one another. If they did, they must eventually
run into one another. It would be like meeting someone else
running away from the police in the opposite direction—you
would both be caught! (Or, in this case, fall into a black
hole.) But if these light rays were swallowed up by the black
hole, then they could not have been on the boundary of the
black hole. So the paths of light rays in the event horizon had
always to be moving parallel to, or away from, each other.
Another way of seeing this is that the event horizon, the
boundary of the black hole, is like the edge of a shadow—the
shadow of impending doom. If you look at the shadow cast
by a source at a great distance, such as the sun, you will see
that the rays of light in the edge are not approaching each
other.

If the rays of light that form the event horizon, the bound-

FIGURE 7.2 AND FIGURE 7.3

ary of the black hole, can never approach each other, the area of the event horizon might stay the same or increase with time but it could never decrease—because that would mean that at least some of the rays of light in the boundary would have to be approaching each other. In fact, the area would increase whenever matter or radiation fell into the black hole (Fig. 7.2). Or if two black holes collided and merged together to form a single black hole, the area of the event horizon of the final black hole would be greater than or equal to the sum of the areas of the event horizons of the original black holes (Fig. 7.3). This nondecreasing property of the event horizon's area placed an important restriction on the possible behavior of black holes. I was so excited with my discovery that I did not get much sleep that night. The next day I rang up Roger Penrose. He agreed with me. I think, in fact, that he had been aware of this property of the area. However, he had been using a slightly different definition of a black hole. He had not realized that the boundaries of the black hole according to the two definitions would be the same, and hence so would their areas, provided the black hole had settled down to a state in which it was not changing with time.

The nondecreasing behavior of a black hole's area was very reminiscent of the behavior of a physical quantity called entropy, which measures the degree of disorder of a system. It is a matter of common experience that disorder will tend to increase if things are left to themselves. (One has only to stop making repairs around the house to see that!) One can create order out of disorder (for example, one can paint the house), but that requires expenditure of effort or energy and so decreases the amount of ordered energy available.

A precise statement of this idea is known as the second law of thermodynamics. It states that the entropy of an isolated system always increases, and that when two systems are joined together, the entropy of the combined system is greater than the sum of the entropies of the individual systems. For example, consider a system of gas molecules in a box. The molecules can be thought of as little billiard balls continually colliding with each other and bouncing off the walls of the box. The higher the temperature of the gas, the faster the molecules move, and so the more frequently and harder they collide with the walls of the box and the greater the outward pressure they exert on the walls. Suppose that initially the molecules are all confined to the left-hand side of the box by a partition. If the partition is then removed, the molecules will tend to spread out and occupy both halves of the box. At some later time they could, by chance, all be in the right half or back in the left half, but it is overwhelmingly more probable that there will be roughly equal numbers in the two halves. Such a state is less ordered, or more disordered, than the original state in which all the molecules were in one half. One therefore says that the entropy of the gas has gone up. Similarly, suppose one starts with two boxes, one containing oxygen molecules and the other containing nitrogen molecules. If one joins the boxes together and removes the intervening wall, the oxygen and the nitrogen molecules will start to mix. At a later time the most probable state would be a fairly uniform mixture of oxygen and nitrogen molecules throughout the two boxes. This state would be less ordered, and hence have more entropy, than the initial state of two separate boxes.

The second law of thermodynamics has a rather different status than that of other laws of science, such as Newton's law of gravity, for example, because it does not hold always, just in the vast majority of cases. The probability of all the gas molecules in our first box being found in one half of the box at a later time is many millions of millions to one, but it can happen. However, if one has a black hole around, there seems to be a rather easier way of violating the second law: just throw some matter with a lot of entropy, such as a box of gas, down the black hole. The total entropy of matter outside the black hole would go down. One could, of course, still say that the total entropy, including the entropy inside the black hole, has not gone down—but since there is no way to look inside the black hole, we cannot see how much entropy the matter inside it has. It would be nice, then, if there was some feature of the black hole by which observers outside the black hole could tell its entropy, and which would increase whenever matter carrying entropy fell into the black hole. Following the discovery, described above, that the area of the event horizon increased whenever matter fell into a black hole, a research student at Princeton named Jacob Bekenstein suggested that the area of the event horizon was a measure of the entropy of the black hole. As matter carrying entropy fell into a black hole, the area of its event horizon would go up, so that the sum of the entropy of matter outside black holes and the area of the horizons would never go down.

This suggestion seemed to prevent the second law of thermodynamics from being violated in most situations. However, there was one fatal flaw. If a black hole has entropy, then it ought also to have a temperature. But a body with a particular temperature must emit radiation at a certain rate. It is a matter of common experience that if one heats up a poker in a fire it glows red hot and emits radiation, but bodies at lower temperatures emit radiation too; one just does not normally notice it because the amount is fairly small. This radiation is required in order to prevent violation of the second law. So black holes ought to emit radiation. But by their very definition, black holes are objects that are not supposed to emit anything. It therefore seemed that the area of the event horizon of a black hole could not be regarded as its entropy. In 1972 I wrote a paper with Brandon Carter and an American colleague, Jim Bardeen, in which we pointed out that although there were many similarities between entropy and the area of the event horizon, there was this apparently fatal difficulty. I must admit that in writing this paper I was motivated partly by irritation with Bekenstein, who, I felt, had misused my discovery of the increase of the area of the event horizon. However, it turned out in the end that he was basically correct, though in a manner he had certainly not expected.

In September 1973, while I was visiting Moscow, I discussed black holes with two leading Soviet experts, Yakov Zeldovich and Alexander Starobinsky. They convinced me that, according to the quantum mechanical uncertainty principle, rotating black holes should create and emit particles. I believed their arguments on physical grounds, but I did not like the mathematical way in which they calculated the emission. I therefore set about devising a better mathematical treatment, which I described at an informal seminar in Oxford at the end of November 1973. At that time I had not done the calculations to find out how much would actually be emitted. I was expecting to discover just the radiation that Zeldovich and Starobinsky had predicted from rotating black holes. However, when I did the calculation, I found, to my surprise and annoyance, that even nonrotating black holes should apparently create and emit particles at a steady rate. At first I thought that this emission indicated that one of the approximations I had used was not valid. I was afraid that if Bekenstein found out about it, he would use it as a further argument to support his ideas about the entropy of black holes, which I still did not like. However, the more I thought about it, the more it seemed that the approximations really ought to hold. But what finally convinced me that the emission was real was that the spectrum of the emitted particles was exactly that which would be emitted by a hot body, and that the black hole was emitting particles at exactly the correct rate to prevent violations of the second law. Since then the calculations have been repeated in a number of different forms by other people. They all confirm that a black hole ought to emit particles and radiation as if it were a hot body with a temperature that depends only on the black hole's mass: the higher the mass, the lower the temperature.

How is it possible that a black hole appears to emit particles when we know that nothing can escape from within its event horizon? The answer, quantum theory tells us, is that the particles do not come from within the black hole, but from the "empty" space just outside the black hole's event horizon! We can understand this in the following way: What we think of as "empty" space cannot be completely empty because that would mean that all the fields, such as the gravitational and electromagnetic fields, would have to be exactly zero. However, the value of a field and its rate of change with time are like the position and velocity of a particle: the uncertainty principle implies that the more accurately one knows one of these quantities, the less accurately one can know the other. So in empty space the field cannot be fixed at exactly zero, because then it would have both a precise value (zero) and a precise rate of change (also zero). There must be a certain minimum amount of uncertainty, or quantum fluctuations, in the value of the field. One can think of these fluctuations as pairs of particles of light or gravity that appear together at some time, move apart, and then come together again and annihilate each other. These particles are virtual particles like the particles that carry the gravitational force of the sun: unlike real particles, they cannot be observed directly with a particle detector. However, their indirect effects, such as small changes in the energy of electron orbits in atoms, can be measured and agree with the theoretical predictions to a remarkable degree of accuracy. The uncertainty principle also predicts that there will be similar virtual pairs of matter particles, such as electrons or quarks. In this case, however, one member of the pair will be a particle and the other an antiparticle (the antiparticles of light and gravity are the same as the particles).

Because energy cannot be created out of nothing, one of the partners in a particle/antiparticle pair will have positive energy, and the other partner negative energy. The one with negative energy is condemned to be a short-lived virtual particle because real particles always have positive energy in normal situations. It must therefore seek out its partner and annihilate with it. However, a real particle close to a massive body has less energy than if it were far away, because it would take energy to lift it far away against the gravitational attraction of the body. Normally, the energy of the particle is still positive, but the gravitational field inside a black hole is

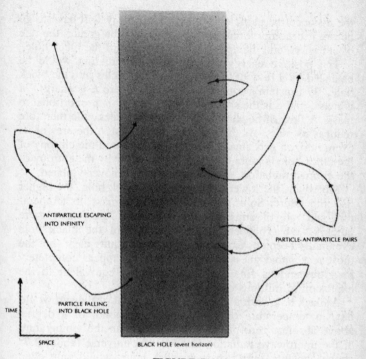

ANTIPARTICLE ESCAPING
INTO INFINITY

PARTICLE-ANTIPARTICLE PAIRS

TIME

PARTICLE FALLING
INTO BLACK HOLE

SPACE

BLACK HOLE (event horizon)

FIGURE 7.4

so strong that even a real particle can have negative energy there. It is therefore possible, if a black hole is present, for the virtual particle with negative energy to fall into the black hole and become a real particle or antiparticle. In this case it no longer has to annihilate with its partner. Its forsaken partner may fall into the black hole as well. Or, having positive energy, it might also escape from the vicinity of the black hole as a real particle or antiparticle (Fig. 7.4). To an observer at a distance, it will appear to have been emitted from the black hole. The smaller the black hole, the shorter the distance the particle with negative energy will have to go before it becomes a real particle, and thus the greater the rate of emission, and the apparent temperature, of the black hole.

The positive energy of the outgoing radiation would be balanced by a flow of negative energy particles into the black hole. By Einstein's equation $E = mc^2$ (where E is energy, m is mass, and c is the speed of light), energy is proportional to mass. A flow of negative energy into the black hole therefore reduces its mass. As the black hole loses mass, the area of its event horizon gets smaller, but this decrease in the entropy of the black hole is more than compensated for by the entropy of the emitted radiation, so the second law is never violated.

Moreover, the lower the mass of the black hole, the higher its temperature. So as the black hole loses mass, its temperature and rate of emission increase, so it loses mass more quickly. What happens when the mass of the black hole eventually becomes extremely small is not quite clear, but the most reasonable guess is that it would disappear completely in a tremendous final burst of emission, equivalent to the explosion of millions of H-bombs.

A black hole with a mass a few times that of the sun would have a temperature of only one ten millionth of a degree above absolute zero. This is much less than the temperature of the microwave radiation that fills the universe (about 2.7° above absolute zero), so such black holes would emit even less than they absorb. If the universe is destined to go on expanding forever, the temperature of the microwave radiation will eventually decrease to less than that of such a black hole, which will then begin to lose mass. But, even then, its temperature would be so low that it would take about a million million million million million million million million million million million years (1 with sixty-six zeros after it) to evaporate completely. This is much longer than the age of the universe, which is only about ten or twenty thousand million years (1 or 2 with ten zeros after it). On the other hand, as mentioned in Chapter 6, there might be primordial black holes with a very much smaller mass that were made by the collapse of irregularities in the very early stages of the universe. Such black holes would have a much higher temperature and would be emitting radiation at a much greater rate. A primordial black hole with an initial mass of a thousand million tons would have a lifetime roughly equal to the age of the universe. Primordial black holes with initial masses less than this figure would already have completely evaporated, but those with slightly greater masses would still be emitting radiation in the form of X rays and gamma rays. These X rays and gamma rays are like waves of light, but with a much shorter wavelength. Such holes hardly deserve the epithet *black:* they really are *white hot* and are emitting energy at a rate of about ten thousand megawatts.

One such black hole could run ten large power stations, if only we could harness its power. This would be rather difficult, however: the black hole would have the mass of a mountain compressed into less than a million millionth of an inch, the size of the nucleus of an atom! If you had one of these black holes on the surface of the earth, there would be no way to stop it from falling through the floor to the center of the earth. It would oscillate through the earth and back, until eventually it settled down at the center. So the only place to put such a black hole, in which one might use the energy that it emitted, would be in orbit around the earth—and the only way that one could get it to orbit the earth would be to attract it there by towing a large mass in front of it, rather like a carrot in front of a donkey. This does not sound like a very practical proposition, at least not in the immediate future.

But even if we cannot harness the emission from these primordial black holes, what are our chances of observing them? We could look for the gamma rays that the primordial black holes emit during most of their lifetime. Although the radiation from most would be very weak because they are far away, the total from all of them might be detectable. We do observe such a background of gamma rays: Fig. 7.5 shows how the observed intensity differs at different frequencies (the

FIGURE 7.5

number of waves per second). However, this background could have been, and probably was, generated by processes other than primordial black holes. The dotted line in Fig. 7.5 shows how the intensity should vary with frequency for gamma rays given off by primordial black holes, if there were on average 300 per cubic light-year. One can therefore say that the observations of the gamma ray background do not provide any *positive* evidence for primordial black holes, but they do tell us that on average there cannot be more than 300 in every cubic light-year in the universe. This limit means that primordial black holes could make up at most one millionth of the matter in the universe.

With primordial black holes being so scarce, it might seem unlikely that there would be one near enough for us to observe as an individual source of gamma rays. But since gravity would draw primordial black holes toward any matter, they should be much more common in and around galaxies. So although the gamma ray background tells us that there can be no more than 300 primordial black holes per cubic light-year on average, it tells us nothing about how common they might be in our own galaxy. If they were, say, a million times more common than this, then the nearest black hole to us would probably be at a distance of about a thousand million kilometers, or about as far away as Pluto, the farthest known planet. At this distance it would still be very difficult to detect the steady emission of a black hole, even if it was ten thousand megawatts. In order to observe a primordial black hole one would have to detect several gamma ray quanta coming from the same direction within a reasonable space of time, such as a week. Otherwise, they might simply be part of the background. But Planck's quantum principle tells us that each gamma ray quantum has a very high energy, because gamma rays have a very high frequency, so it would not take many quanta to radiate even ten thousand megawatts. And to observe these few coming from the distance of Pluto would require a larger gamma ray detector than any that have been constructed so far. Moreover, the detector would have to be in space, because gamma rays cannot penetrate the atmosphere.

Of course, if a black hole as close as Pluto were to reach the end of its life and blow up, it would be easy to detect the final burst of emission. But if the black hole has been emitting for the last ten or twenty thousand million years, the chance of it reaching the end of its life within the next few years, rather than several million years in the past or future, is really rather small! So in order to have a reasonable chance of seeing an explosion before your research grant ran out, you would have to find a way to detect any explosions within a distance of about one light-year. You would still have the problem of needing a large gamma ray detector to observe several gamma ray quanta from the explosion. However, in this case, it would not be necessary to determine that all the quanta came from the same direction: it would be enough to observe that they all arrived within a very short time interval to be reasonably confident that they were coming from the same burst.

One gamma ray detector that might be capable of spotting primordial black holes is the entire earth's atmosphere. (We are, in any case, unlikely to be able to build a larger detector!) When a high-energy gamma ray quantum hits the atoms in our atmosphere, it creates pairs of electrons and positrons (antielectrons). When these hit other atoms they in turn create more pairs of electrons and positrons, so one gets what is called an electron shower. The result is a form of light called Cerenkov radiation. One can therefore detect gamma ray bursts by looking for flashes of light in the night sky. Of course, there are a number of other phenomena, such as lightning and reflections of sunlight off tumbling satellites and orbiting debris, that can also give flashes in the sky. One could distinguish gamma ray bursts from such effects by observing flashes simultaneously at two or more fairly widely separated locations. A search like this has been carried out by two scientists from Dublin, Neil Porter and Trevor Weekes, using telescopes in Arizona. They found a number of flashes but none that could be definitely ascribed to gamma ray bursts from primordial black holes.

Even if the search for primordial black holes proves negative, as it seems it may, it will still give us important information about the very early stages of the universe. If the early universe had been chaotic or irregular, or if the pressure of matter had been low, one would have expected it to produce many more primordial black holes than the limit already set by our observations of the gamma ray background. Only if the early universe was very smooth and uniform, with a high pressure, can one explain the absence of observable numbers of primordial black holes.

*   *   *

The idea of radiation from black holes was the first example of a prediction that depended in an essential way on both the great theories of this century, general relativity and quantum mechanics. It aroused a lot of opposition initially because it upset the existing viewpoint: "How can a black hole emit anything?" When I first announced the results of my calculations at a conference at the Rutherford-Appleton Laboratory near Oxford, I was greeted with general incredulity. At the end of my talk the chairman of the session, John G. Taylor from Kings College, London, claimed it was all nonsense. He even wrote a paper to that effect. However, in the end most people, including John Taylor, have come to the conclusion that black holes must radiate like hot bodies if our other ideas about general relativity and quantum mechanics are correct. Thus even though we have not yet managed to find a primordial black hole, there is fairly general agreement that if we did, it would have to be emitting a lot of gamma rays and X rays.

The existence of radiation from black holes seems to imply that gravitational collapse is not as final and irreversible as we once thought. If an astronaut falls into a black hole, its mass will increase, but eventually the energy equivalent of that extra mass will be returned to the universe in the form of radiation. Thus, in a sense, the astronaut will be "recycled." It would be a poor sort of immortality, however, because any personal concept of time for the astronaut would almost certainly come to an end as he was torn apart inside the black hole! Even the types of particles that were eventually emitted by the black hole would in general be different from those that made up the astronaut: the only feature of the astronaut that would survive would be his mass or energy.

The approximations I used to derive the emission from black holes should work well when the black hole has a mass greater than a fraction of a gram. However, they will break down at the end of the black hole's life when its mass gets very small. The most likely outcome seems to be that the black hole will just disappear, at least from our region of the universe, taking with it the astronaut and any singularity there might be inside it, if indeed there is one. This was the first indication that quantum mechanics might remove the singularities that were predicted by general relativity. However, the methods that I and other people were using in 1974 were not able to answer questions such as whether singularities would occur in quantum gravity. From 1975 onward I therefore started to develop a more powerful approach to quantum gravity based on Richard Feynman's idea of a sum over histories. The answers that this approach suggests for the origin and fate of the universe and its contents, such as astronauts, will be described in the next two chapters. We shall see that although the uncertainty principle places limitations on the accuracy of all our predictions, it may at the same time remove the fundamental unpredictability that occurs at a space-time singularity.

8

The Origin
and Fate of
the Universe

Einstein's general theory of relativity, on its own, predicted that space-time began at the big bang singularity and would come to an end either at the big crunch singularity (if the whole universe recollapsed), or at a singularity inside a black hole (if a local region, such as a star, were to collapse). Any matter that fell into the hole would be destroyed at the singularity, and only the gravitational effect of its mass would continue to be felt outside. On the other hand, when quantum effects were taken into account, it seemed that the mass or energy of the matter would eventually be returned to the rest of the universe, and that the black hole, along with any singularity inside it, would evaporate away and finally disappear. Could quantum mechanics have an equally dramatic effect on the big bang and big crunch singularities? What really happens during the very early or late stages of the universe, when gravitational fields are so strong that quantum effects cannot be ignored? Does the universe in fact have a beginning or an end? And if so, what are they like?

Throughout the 1970s I had been mainly studying black holes, but in 1981 my interest in questions about the origin and fate of the universe was reawakened when I attended a conference on cosmology organized by the Jesuits in the Vatican. The Catholic Church had made a bad mistake with Galileo when it tried to lay down the law on a question of science, declaring that the sun went round the earth. Now, centuries later, it had decided to invite a number of experts to advise it on cosmology. At the end of the conference the participants were granted an audience with the pope. He told us that it was all right to study the evolution of the universe after the big bang, but we should not inquire into the big bang itself because that was the moment of Creation and therefore the work of God. I was glad then that he did not know the subject of the talk I had just given at the conference—the possibility that space-time was finite but had no boundary, which means that it had no beginning, no moment of Creation. I had no desire to share the fate of Galileo, with whom I feel a strong sense of identity, partly because of the coincidence of having been born exactly 300 years after his death!

In order to explain the ideas that I and other people have had about how quantum mechanics may affect the origin and fate of the universe, it is necessary first to understand the generally accepted history of the universe, according to what is known as the "hot big bang model." This assumes that the universe is described by a Friedmann model, right back to the big bang. In such models one finds that as the universe expands, any matter or radiation in it gets cooler. (When the universe doubles in size, its temperature falls by half.) Since temperature is simply a measure of the average energy—or speed—of the particles, this cooling of the universe would have a major effect on the matter in it. At very high temperatures, particles would be moving around so fast that they could escape any attraction toward each other due to nuclear or electromagnetic forces, but as they cooled off one would expect particles that attract each other to start to clump together. Moreover, even the types of particles that exist in the universe would depend on the temperature. At high enough temperatures, particles have so much energy that whenever they collide many different particle/antiparticle pairs would be produced—and although some of these particles would annihilate on hitting antiparticles, they would be produced more rapidly than they could annihilate. At lower temperatures, however, when colliding particles have less energy, particle/antiparticle pairs would be produced less quickly— and annihilation would become faster than production.

At the big bang itself, the universe is thought to have had zero size, and so to have been infinitely hot. But as the universe expanded, the temperature of the radiation decreased. One second after the big bang, it would have fallen to about ten thousand million degrees. This is about a thousand times the temperature at the center of the sun, but temperatures as high as this are reached in H-bomb explosions. At this time the universe would have contained mostly photons, electrons, and neutrinos (extremely light particles that are affected only by the weak force and gravity) and their antiparticles, together with some protons and neutrons. As the universe continued to expand and the temperature to drop, the rate at which electron/antielectron pairs were being produced in collisions would have fallen below the rate at which they were being destroyed by annihilation. So most of the electrons and antielectrons would have annihilated with each other to produce more photons, leaving only a few electrons left over. The neutrinos and antineutrinos, however, would not have annihilated with each other, because these particles interact with themselves and with other particles only very weakly. So they should still be around today. If we could observe them, it would provide a good test of this picture of a very hot early stage of the universe. Unfortunately, their energies nowadays would be too low for us to observe them directly. However, if neutrinos are not massless, but have a small mass of their own, as suggested by an unconfirmed Russian experiment performed in 1981, we might be able to detect them indirectly: they could be a form of "dark matter," like that mentioned earlier, with sufficient gravitational attraction to stop the expansion of the universe and cause it to collapse again.

About one hundred seconds after the big bang, the temperature would have fallen to one thousand million degrees, the temperature inside the hottest stars. At this temperature protons and neutrons would no longer have sufficient energy to escape the attraction of the strong nuclear force, and would have started to combine together to produce the nuclei of atoms of deuterium (heavy hydrogen), which contain one proton and one neutron. The deuterium nuclei then would have combined with more protons and neutrons to make helium nuclei, which contain two protons and two neutrons, and also small amounts of a couple of heavier elements, lithium and beryllium. One can calculate that in the hot big bang model about a quarter of the protons and neutrons would have been converted into helium nuclei, along with a small amount of heavy hydrogen and other elements. The remaining neutrons would have decayed into protons, which are the nuclei of ordinary hydrogen atoms.

This picture of a hot early stage of the universe was first put forward by the scientist George Gamow in a famous paper written in 1948 with a student of his, Ralph Alpher. Gamow had quite a sense of humor—he persuaded the nuclear scientist Hans Bethe to add his name to the paper to make the list of authors "Alpher, Bethe, Gamow," like the first three letters of the Greek alphabet, alpha, beta, gamma: particularly appropriate for a paper on the beginning of the universe! In this paper they made the remarkable prediction that radiation (in the form of photons) from the very hot early stages of the universe should still be around today, but with its temperature reduced to only a few degrees above absolute zero (−273° C). It was this radiation that Penzias and Wilson found in 1965. At the time that Alpher, Bethe, and Gamow wrote their paper, not much was known about the nuclear reactions of protons and neutrons. Predictions made for the proportions of various elements in the early universe were therefore rather inaccurate, but these calculations have been repeated in the light of better knowledge and now agree very well with what we observe. It is, moreover, very difficult to explain in any other way why there should be so much helium in the universe. We are therefore fairly confident that we have the right picture, at least back to about one second after the big bang.

Within only a few hours of the big bang, the production of helium and other elements would have stopped. And after that, for the next million years or so, the universe would have just continued expanding, without anything much happening. Eventually, once the temperature had dropped to a few thousand degrees, and electrons and nuclei no longer had enough energy to overcome the electromagnetic attraction between them, they would have started combining to form atoms. The universe as a whole would have continued expanding and cooling, but in regions that were slightly denser than average, the expansion would have been slowed down by the extra gravitational attraction. This would eventually stop expansion in some regions and cause them to start to recollapse. As they were collapsing, the gravitational pull of matter outside these regions might start them rotating slightly. As the collapsing region got smaller, it would spin faster—just as skaters spinning on ice spin faster as they draw in their arms. Eventually, when the region got small enough, it would be spinning fast enough to balance the attraction of gravity, and in this way disklike rotating galaxies were born. Other regions, which did not happen to pick up a rotation, would become oval-shaped objects called elliptical galaxies. In these, the region would stop collapsing because individual parts of the galaxy would be orbiting stably round its center, but the galaxy would have no overall rotation.

As time went on, the hydrogen and helium gas in the galaxies would break up into smaller clouds that would collapse under their own gravity. As these contracted, and the atoms within them collided with one another, the temperature of the gas would increase, until eventually it became hot enough to start nuclear fusion reactions. These would convert the hydrogen into more helium, and the heat given off would raise the pressure, and so stop the clouds from contracting any further. They would remain stable in this state for a long time as stars like our sun, burning hydrogen into helium and radiating the resulting energy as heat and light. More massive stars would need to be hotter to balance their stronger gravitational attraction, making the nuclear fusion reactions proceed so much more rapidly that they would use up their hydrogen in as little as a hundred million years. They would then contract slightly, and as they heated up further, would start to convert helium into heavier elements like carbon or oxygen. This, however, would not release much more energy, so a crisis would occur, as was described in the chapter on black holes. What happens next is not completely clear, but it seems likely that the central regions of the star would collapse to a very dense state, such as a neutron star or black hole. The outer regions of the star may sometimes get blown off in a tremendous explosion called a supernova, which would outshine all the other stars in its galaxy. Some of the heavier elements produced near the end of the star's life would be flung back into the gas in the galaxy, and would provide some of the raw material for the next generation of stars. Our own sun contains about 2 percent of these heavier elements because it is a second- or third-generation star, formed some five thousand million years ago out of a cloud of rotating gas containing the debris of earlier supernovas. Most of the gas in that cloud went to form the sun or got blown away, but a small amount of the heavier elements collected together to form the bodies that now orbit the sun as planets like the earth.

The earth was initially very hot and without an atmosphere. In the course of time it cooled and acquired an atmosphere from the emission of gases from the rocks. This early atmosphere was not one in which we could have survived. It contained no oxygen, but a lot of other gases that are poisonous to us, such as hydrogen sulfide (the gas that gives rotten eggs their smell). There are, however, other primitive forms of life that can flourish under such conditions. It is thought that they developed in the oceans, possibly as a result of chance combinations of atoms into large structures, called macromolecules, which were capable of assembling other atoms in the ocean into similar structures. They would thus have reproduced themselves and multiplied. In some cases there would be errors in the reproduction. Mostly these errors would have been such that the new macromolecule could not reproduce itself and eventually would have been destroyed. However, a few of the errors would have produced new macromolecules that were even better at reproducing themselves. They would have therefore had an advantage and would have tended to replace the original macromolecules. In this way a process of evolution was started that led to the development of more and more complicated, self-reproducing organisms. The first primitive forms of life consumed various materials, including hydrogen sulfide, and released oxygen. This gradually changed the atmosphere to the composition that it has today and allowed the development of higher forms of life such as fish, reptiles, mammals, and ultimately the human race.

This picture of a universe that started off very hot and cooled as it expanded is in agreement with all the observational evidence that we have today. Nevertheless, it leaves a number of important questions unanswered:

(1) Why was the early universe so hot?

(2) Why is the universe so uniform on a large scale? Why does it look the same at all points of space and in all directions? In particular, why is the temperature of the microwave background radiation so nearly the same when we look in different directions? It is a bit like asking a number of students an exam question. If they all give exactly the same answer, you can be pretty sure they

have communicated with each other. Yet, in the model described above, there would not have been time since the big bang for light to get from one distant region to another, even though the regions were close together in the early universe. According to the theory of relativity, if light cannot get from one region to another, no other information can. So there would be no way in which different regions in the early universe could have come to have had the same temperature as each other, unless for some unexplained reason they happened to start out with the same temperature.

(3) Why did the universe start out with so nearly the critical rate of expansion that separates models that recollapse from those that go on expanding forever, so that even now, ten thousand million years later, it is still expanding at nearly the critical rate? If the rate of expansion one second after the big bang had been smaller by even one part in a hundred thousand million million, the universe would have recollapsed before it ever reached its present size.

(4) Despite the fact that the universe is so uniform and homogeneous on a large scale, it contains local irregularities, such as stars and galaxies. These are thought to have developed from small differences in the density of the early universe from one region to another. What was the origin of these density fluctuations?

The general theory of relativity, on its own, cannot explain these features or answer these questions because of its prediction that the universe started off with infinite density at the big bang singularity. At the singularity, general relativity and all other physical laws would break down: one couldn't predict what will come out of the singularity. As explained before, this means that one might as well cut the big bang, and any events before it, out of the theory, because they can have no effect on what we observe. Space-time *would* have a boundary—a beginning at the big bang.

Science seems to have uncovered a set of laws that, within the limits set by the uncertainty principle, tell us how the universe will develop with time, if we know its state at any one time. These laws may have originally been decreed by God, but it appears that he has since left the universe to evolve according to them and does not now intervene in it. But how did he choose the initial state or configuration of the universe? What were the "boundary conditions" at the beginning of time?

One possible answer is to say that God chose the initial configuration of the universe for reasons that we cannot hope to understand. This would certainly have been within the power of an omnipotent being, but if he had started it off in such an incomprehensible way, why did he choose to let it evolve according to laws that we could understand? The whole history of science has been the gradual realization that events do not happen in an arbitrary manner, but that they reflect a certain underlying order, which may or may not be divinely inspired. It would be only natural to suppose that this order should apply not only to the laws, but also to the conditions at the boundary of space-time that specify the initial state of the universe. There may be a large number of models of the universe with different initial conditions that all obey the laws. There ought to be some principle that picks out one initial state, and hence one model, to represent our universe.

One such possibility is what are called chaotic boundary conditions. These implicitly assume either that the universe is spatially infinite or that there are infinitely many universes. Under chaotic boundary conditions, the probability of finding any particular region of space in any given configuration just after the big bang is the same, in some sense, as the probability of finding it in any other configuration: the initial state of the universe is chosen purely randomly. This would mean that the early universe would have probably been very chaotic and irregular because there are many more chaotic and disordered configurations for the universe than there are smooth and ordered ones. (If each configuration is equally probable, it is likely that the universe started out in a chaotic and disordered state, simply because there are so many more of them.) It is difficult to see how such chaotic initial conditions could have given rise to a universe that is so smooth and regular on a large scale as ours is today. One would also have expected the density fluctuations in such a model to have led to the formation of many more primordial black holes than the upper limit that has been set by observations of the gamma ray background.

If the universe is indeed spatially infinite, or if there are infinitely many universes, there would probably be some large regions somewhere that started out in a smooth and uniform manner. It is a bit like the well-known horde of monkeys hammering away on typewriters—most of what they write will be garbage, but very occasionally by pure chance they will type out one of Shakespeare's sonnets. Similarly, in the case of the universe, could it be that we are living in a region that just happens by chance to be smooth and uniform? At first sight this might seem very improbable, because such smooth regions would be heavily outnumbered by chaotic and irregular regions. However, suppose that only in the smooth regions were galaxies and stars formed and were conditions right for the development of complicated self-replicating organisms like ourselves who were capable of asking the question: Why is the universe so smooth? This is an example of the application of what is known as the anthropic principle, which can be paraphrased as "We see the universe the way it is because we exist."

There are two versions of the anthropic principle, the weak and the strong. The weak anthropic principle states that in a universe that is large or infinite in space and/or time, the conditions necessary for the development of intelligent life will be met only in certain regions that are limited in space and time. The intelligent beings in these regions should therefore not be surprised if they observe that their locality in the universe satisfies the conditions that are necessary for their existence. It is a bit like a rich person living in a wealthy neighborhood not seeing any poverty.

One example of the use of the weak anthropic principle is to "explain" why the big bang occurred about ten thousand million years ago—it takes about that long for intelligent beings to evolve. As explained above, an early generation of stars first had to form. These stars converted some of the original hydrogen and helium into elements like carbon and oxygen, out of which we are made. The stars then exploded as supernovas, and their debris went to form other stars and planets, among them those of our Solar System, which is about five thousand million years old. The first one or two thousand million years of the earth's existence were too hot for the development of anything complicated. The remaining three thousand million years or so have been taken up by the slow process of biological evolution, which has led from the simplest organisms to beings who are capable of measuring time back to the big bang.

Few people would quarrel with the validity or utility of the weak anthropic principle. Some, however, go much further and propose a strong version of the principle. According to this theory, there are either many different universes or many different regions of a single universe, each with its own initial configuration and, perhaps, with its own set of laws of science. In most of these universes the conditions would not be right for the development of complicated organisms; only in the few universes that are like ours would intelligent beings develop and ask the question: "Why is the universe the way we see it?" The answer is then simple: If it had been different, we would not be here!

The laws of science, as we know them at present, contain many fundamental numbers, like the size of the electric charge of the electron and the ratio of the masses of the proton and the electron. We cannot, at the moment at least, predict the values of these numbers from theory—we have to find them by observation. It may be that one day we shall discover a complete unified theory that predicts them all, but it is also possible that some or all of them vary from universe to universe or within a single universe. The remarkable fact is that the values of these numbers seem to have been very finely adjusted to make possible the development of life. For example if the electric charge of the electron had been only slightly different, stars either would have been unable to burn hydrogen and helium, or else they would not have exploded. Of course, there might be other forms of intelligent life, not dreamed of even by writers of science fiction, that did not require the light of a star like the sun or the heavier chemical elements that are made in stars and are flung back into space when the stars explode. Nevertheless, it seems clear that there are relatively few ranges of values for the numbers that would allow the development of any form of intelligent life. Most sets of values would give rise to universes that, although they might be very beautiful, would contain no one able to wonder at that beauty. One can take this either as evidence of a divine purpose in Creation and the choice of the laws of science or as support for the strong anthropic principle.

There are a number of objections that one can raise to the strong anthropic principle as an explanation of the observed state of the universe. First, in what sense can all these different universes be said to exist? If they are really separate from each other, what happens in another universe can have no observable consequences in our own universe. We should therefore use the principle of economy and cut them out of the theory. If, on the other hand, they are just different regions of a single universe, the laws of science would have to be the same in each region, because otherwise one could not move continuously from one region to another. In this case the only difference between the regions would be their initial configurations and so the strong anthropic principle would reduce to the weak one.

A second objection to the strong anthropic principle is that it runs against the tide of the whole history of science. We have developed from the geocentric cosmologies of Ptolemy and his forebears, through the heliocentric cosmology of Copernicus and Galileo, to the modern picture in which the earth is a medium-sized planet orbiting around an average star in the outer suburbs of an ordinary spiral galaxy, which is itself only one of about a million million galaxies in the observable universe. Yet the strong anthropic principle would claim that this whole vast construction exists simply for our sake. This is very hard to believe. Our Solar System is certainly a prerequisite for our existence, and one might extend this to the whole of our galaxy to allow for an earlier generation of stars that created the heavier elements. But there does not seem to be any need for all those other galaxies, nor for the universe to be so uniform and similar in every direction on the large scale.

One would feel happier about the anthropic principle, at least in its weak version, if one could show that quite a number of different initial configurations for the universe would have evolved to produce a universe like the one we observe. If this is the case, a universe that developed from some sort of random initial conditions should contain a number of regions that are smooth and uniform and are suitable for the evolution of intelligent life. On the other hand, if the initial state of the universe had to be chosen extremely carefully to lead to something like what we see around us, the universe would be unlikely to contain *any* region in which life would appear. In the hot big bang model described above, there was not enough time in the early universe for heat to have flowed from one region to another. This means that the initial state of the universe would have to have had exactly the same temperature everywhere in order to account for the fact that the microwave background has the same temperature in every direction we look. The initial rate of expansion also would have had to be chosen very precisely for the rate of expansion still to be so close to the critical rate needed to avoid recollapse. This means that the initial state of the universe must have been very carefully chosen indeed if the hot big bang model was correct right back to the beginning of time. It would be very difficult to explain why the universe should have begun in just this way, except as the act of a God who intended to create beings like us.

In an attempt to find a model of the universe in which many different initial configurations could have evolved to something like the present universe, a scientist at the Massachusetts Institute of Technology, Alan Guth, suggested that the early universe might have gone through a period of very rapid expansion. This expansion is said to be "inflationary," meaning that the universe at one time expanded at an increasing rate rather than the decreasing rate that it does today. According to Guth, the radius of the universe increased by a million million million million million (1 with thirty zeros after it) times in only a tiny fraction of a second.

Guth suggested that the universe started out from the big bang in a very hot, but rather chaotic, state. These high temperatures would have meant that the particles in the universe would be moving very fast and would have high energies. As we discussed earlier, one would expect that at such high temperatures the strong and weak nuclear forces and the electromagnetic force would all be unified into a single force. As the universe expanded, it would cool, and particle energies would go down. Eventually there would be what is called a phase transition and the symmetry between the forces would be broken: the strong force would become different from the weak and electromagnetic forces. One common example of a phase transition is the freezing of water when you cool it down. Liquid water is symmetrical, the same at every point and in every direction. However, when ice crystals form, they will have definite positions and will be lined up in some direction. This breaks water's symmetry.

In the case of water, if one is careful, one can "supercool" it, that is, one can reduce the temperature below the freezing point (0°C) without ice forming. Guth suggested that the universe might behave in a similar way: the temperature might drop below the critical value without the symmetry between the forces being broken. If this happened, the universe would be in an unstable state, with more energy than if the symmetry had been broken. This special extra energy can be shown to have an antigravitational effect: it would have acted just like the cosmological constant that Einstein introduced into general relativity when he was trying to construct a static model of the universe. Since the universe would already be expanding just as in the hot big bang model, the repulsive effect of this cosmological constant would therefore have made the universe expand at an ever-increasing rate. Even in regions where there were more matter particles than average, the gravitational attraction of the matter would have been outweighed by the repulsion of the effective cosmological constant. Thus these regions would also expand in an accelerating inflationary manner. As they expanded and the matter particles got farther apart, one would be left with an expanding universe that contained hardly any particles and was still in the supercooled state. Any irregularities in the universe would simply have been smoothed out by the expansion, as the wrinkles in a balloon are smoothed away when you blow it up. Thus the present smooth and uniform state of the universe could have evolved from many different nonuniform initial states.

In such a universe, in which the expansion was accelerated by a cosmological constant rather than slowed down by the gravitational attraction of matter, there would be enough time for light to travel from one region to another in the early universe. This could provide a solution to the problem, raised earlier, of why different regions in the early universe have the same properties. Moreover, the rate of expansion of the universe would automatically become very close to the critical rate determined by the energy density of the universe. This could then explain why the rate of expansion is still so close to the critical rate, without having to assume that the initial rate of expansion of the universe was very carefully chosen.

The idea of inflation could also explain why there is so much matter in the universe. There are something like ten million million million million million million million million million million million million million million (1 with eighty zeroes after it) particles in the region of the universe that we can observe. Where did they all come from? The answer is that, in quantum theory, particles can be created out of energy in the form of particle/antiparticle pairs. But that just raises the question of where the energy came from. The answer is that the total energy of the universe is exactly zero. The matter in the universe is made out of positive energy. However, the matter is all attracting itself by gravity. Two pieces of matter that are close to each other have less energy than the same two pieces a long way apart, because you have to expend energy to separate them against the gravitational force that is pulling them together. Thus, in a sense, the gravitational field has negative energy. In the case of a universe that is approximately uniform in space, one can show that this negative gravitational energy exactly cancels the positive energy represented by the matter. So the total energy of the universe is zero.

Now twice zero is also zero. Thus the universe can double the amount of positive matter energy and also double the negative gravitational energy without violation of the conservation of energy. This does not happen in the normal expansion of the universe in which the matter energy density goes down as the universe gets bigger. It does happen, however, in the inflationary expansion because the energy density of the supercooled state remains constant while the universe expands: when the universe doubles in size, the positive matter energy and the negative gravitational energy both double, so the total energy remains zero. During the inflationary phase, the universe increases its size by a very large amount. Thus the total amount of energy available to make particles becomes very large. As Guth has remarked, "It is said that there's no such thing as a free lunch. But the universe is the ultimate free lunch."

The universe is not expanding in an inflationary way today. Thus there has to be some mechanism that would eliminate the very large effective cosmological constant and so change the rate of expansion from an accelerated one to one that is slowed down by gravity, as we have today. In the inflationary expansion one might expect that eventually the symmetry between the forces would be broken, just as supercooled water always freezes in the end. The extra energy of the unbroken symmetry state would then be released and would reheat the universe to a temperature just below the critical temperature for symmetry between the forces. The universe would then go on to expand and cool just like the hot big bang model, but there would now be an explanation of why the universe was expanding at exactly the critical rate and why different regions had the same temperature.

In Guth's original proposal the phase transition was supposed to occur suddenly, rather like the appearance of ice crystals in very cold water. The idea was that "bubbles" of the new phase of broken symmetry would have formed in the old phase, like bubbles of steam surrounded by boiling water. The bubbles were supposed to expand and meet up with each other until the whole universe was in the new phase. The trouble was, as I and several other people pointed out, the universe was expanding so fast that even if the bubbles grew at the speed of light, they would be moving away from each other and so could not join up. The universe would be left in a very nonuniform state, with some regions still having symmetry between the different forces. Such a model of the universe would not correspond to what we see.

In October 1981, I went to Moscow for a conference on quantum gravity. After the conference I gave a seminar on the inflationary model and its problems at the Sternberg Astronomical Institute. Before this, I had got someone else to give my lectures for me, because most people could not understand my voice. But there was not time to prepare this seminar, so I gave it myself, with one of my graduate students repeating my words. It worked well, and gave me much more contact with my audience. In the audience was a young Russian, Andrei Linde, from the Lebedev Institute in Moscow. He said that the difficulty with the bubbles not joining up could be avoided if the bubbles were so big that our region of the universe is all contained inside a single bubble. In order for this to work, the change from symmetry to broken symmetry must have taken place very slowly inside the bubble, but this is quite possible according to grand unified theories. Linde's idea of a slow breaking of symmetry was very good, but I later realized that his bubbles would have to have been bigger than the size of the universe at the time! I showed that instead the symmetry would have broken everywhere at the same time, rather than just inside bubbles. This would lead to a uniform universe, as we observe. I was very excited by this idea and discussed it with one of my students, Ian Moss. As a friend of Linde's, I was rather embarrassed, however, when I was later sent his paper by a scientific journal and asked whether it was suitable for publication. I replied that there was this flaw about the bubbles being bigger than the universe, but that the basic idea of a slow breaking of symmetry was very good. I recommended that the paper be published as it was because it would take Linde several months to correct it, since anything he sent to the West would have to be passed by Soviet censorship, which was neither very skillful nor very quick with scientific papers. Instead, I wrote a short paper with Ian Moss in the same journal in which we pointed out this problem with the bubble and showed how it could be resolved.

The day after I got back from Moscow, I set out for Philadelphia, where I was due to receive a medal from the Franklin Institute. My secretary, Judy Fella, had used her not inconsiderable charm to persuade British Airways to give herself and me free seats on a Concorde as a publicity venture. However, I was held up on my way to the airport by heavy rain and I missed the plane. Nevertheless, I got to Philadelphia in the end and received my medal. I was then asked to give a seminar on the inflationary universe at Drexel University in Philadelphia. I gave the same seminar about the problems of the inflationary universe, just as in Moscow.

A very similar idea to Linde's was put forth independently a few months later by Paul Steinhardt and Andreas Albrecht of the University of Pennsylvania. They are now given joint credit with Linde for what is called "the new inflationary model," based on the idea of a slow breaking of symmetry. (The old inflationary model was Guth's original suggestion of fast symmetry breaking with the formation of bubbles.)

The new inflationary model was a good attempt to explain why the universe is the way it is. However, I and several other people showed that, at least in its original form, it predicted much greater variations in the temperature of the microwave background radiation than are observed. Later work has also cast doubt on whether there could be a phase transition in the very early universe of the kind required. In my personal opinion, the new inflationary model is now dead as a scientific theory, although a lot of people do not seem to have heard of its demise and are still writing papers as if it were viable. A better model, called the chaotic inflationary model, was put forward by Linde in 1983. In this there is no phase transition or supercooling. Instead, there is a spin 0 field, which, because of quantum fluctuations, would have large values in some regions of the early universe. The energy of the field in those regions would behave like a cosmological constant. It would have a repulsive gravitational effect, and thus make those regions expand in an inflationary manner. As they expanded, the energy of the field in them would slowly decrease until the inflationary expansion changed to an expansion like that in the hot big bang model. One of these regions would become what we now see as the observable universe. This model has all the advantages of the earlier inflationary models, but it does not depend on a dubious phase transition, and it can moreover give a reasonable size for the fluctuations in the temperature of the microwave background that agrees with observation.

This work on inflationary models showed that the present state of the universe could have arisen from quite a large number of different initial configurations. This is important, because it shows that the initial state of the part of the universe that we inhabit did not have to be chosen with great care. So we may, if we wish, use the weak anthropic principle to explain why the universe looks the way it does now. It cannot be the case, however, that *every* initial configuration would have led to a universe like the one we observe. One can show this by considering a very different state for the universe at the present time, say, a very lumpy and irregular one. One could use the laws of science to evolve the universe back in time to determine its configuration at earlier times. According to the singularity theorems of classical general relativity, there would still have been a big bang singularity. If you evolve such a universe forward in time according to the laws of science, you will end up with the lumpy and irregular state you started with. Thus there must have been initial configurations that would not have given rise to a universe like the one we see today. So even the inflationary model does not tell us why the initial configuration was not such as to produce something very different from what we observe. Must we turn to the anthropic principle for an explanation? Was it all just a lucky chance? That would seem a counsel of despair, a negation of all our hopes of understanding the underlying order of the universe.

In order to predict how the universe should have started off, one needs laws that hold at the beginning of time. If the classical theory of general relativity was correct, the singularity theorems that Roger Penrose and I proved show that the beginning of time would have been a point of infinite density and infinite curvature of space-time. All the known laws of science would break down at such a point. One might suppose that there were new laws that held at singularities, but it would be very difficult even to formulate such laws at such badly behaved points, and we would have no guide from observations as to what those laws might be. However, what the singularity theorems really indicate is that the gravitational field becomes so strong that quantum gravitational effects become important: classical theory is no longer a good description of the universe. So one has to use a quantum theory of gravity to discuss the very early stages of the universe. As we shall see, it is possible in the quantum theory for the ordinary laws of science to hold everywhere, including at the beginning of time: it is not necessary to postulate new laws for singularities, because there need not be any singularities in the quantum theory.

We don't yet have a complete and consistent theory that combines quantum mechanics and gravity. However, we are fairly certain of some features that such a unified theory should have. One is that it should incorporate Feynman's proposal to formulate quantum theory in terms of a sum over histories. In this approach, a particle does not have just a single history, as it would in a classical theory. Instead, it is supposed to follow every possible path in space-time, and with each of these histories there are associated a couple of numbers, one representing the size of a wave and the other representing its position in the cycle (its phase). The probability that the particle, say, passes through some particular point is found by adding up the waves associated with every possible history that passes through that point. When one actually tries to perform these sums, however, one runs into severe technical problems. The only way around these is the following peculiar prescription: One must add up the waves for particle histories that are not in the "real" time that you and I experience but take place in what is called imaginary time. Imaginary time may sound like science fiction but it is in fact a well-defined mathematical concept. If we take any ordinary (or "real") number and multiply it by itself, the result is a positive number. (For example, 2 times 2 is 4, but so is -2 times -2.) There are, however, special numbers (called imaginary) that give negative numbers when multiplied by themselves. (The one called i, when multiplied by itself, gives $- 1$, $2i$ multiplied by itself gives $- 4$, and so on.) To avoid the technical difficulties with Feynman's sum over histories, one must use imaginary time. That is to say, for the purposes of the calculation one must measure time using imaginary numbers, rather than real ones. This has an interesting effect on space-time: the distinction between time and space disappears completely. A space-time in which events have imaginary values of the time coordinate is said to be Euclidean, after the ancient Greek Euclid, who founded the study of geometry of two-dimensional surfaces. What we now call Euclidean space-time is very similar except that it has four dimensions instead of two. In Euclidean space-time there is no difference between the time direction and directions in space. On the other hand, in real space-time, in which events are labeled by ordinary, real values of the time coordinate, it is easy to tell the difference—the time direction at all points lies within the light cone, and space directions lie outside. In any case, as far as everyday quantum mechanics is concerned, we may regard our use of imaginary time and Euclidean space-time as merely a mathematical device (or trick) to calculate answers about real space-time.

A second feature that we believe must be part of any ultimate theory is Einstein's idea that the gravitational field is represented by curved space-time: particles try to follow the nearest thing to a straight path in a curved space, but because space-time is not flat their paths appear to be bent, as if by a gravitational field. When we apply Feynman's sum over histories to Einstein's view of gravity, the analogue of the history of a particle is now a complete curved space-time that represents the history of the whole universe. To avoid the technical difficulties in actually performing the sum over histories, these curved space-times must be taken to be Euclidean. That is, time is imaginary and is indistinguishable from directions in space. To calculate the probability of finding a real space-time with some certain property, such as looking the same at every point and in every direction, one adds up the waves associated with all the histories that have that property.

In the classical theory of general relativity, there are many different possible curved space-times, each corresponding to a different initial state of the universe. If we knew the initial state of our universe, we would know its entire history. Similarly, in the quantum theory of gravity, there are many different possible quantum states for the universe. Again, if we knew how the Euclidean curved space-times in the sum over histories behaved at early times, we would know the quantum state of the universe.

In the classical theory of gravity, which is based on real space-time, there are only two possible ways the universe can behave: either it has existed for an infinite time, or else it had a beginning at a singularity at some finite time in the past. In the quantum theory of gravity, on the other hand, a third possibility arises. Because one is using Euclidean space-times, in which the time direction is on the same footing as directions in space, it is possible for space-time to be finite in extent and yet to have no singularities that formed a boundary or edge. Space-time would be like the surface of the earth, only with two more dimensions. The surface of the earth is finite in extent but it doesn't have a boundary or edge: if you sail off into the sunset, you don't fall off the edge or run into a singularity. (I know, because I have been round the world!)

If Euclidean space-time stretches back to infinite imaginary time, or else starts at a singularity in imaginary time, we have the same problem as in the classical theory of specifying the initial state of the universe: God may know how the universe began, but we cannot give any particular reason for thinking it began one way rather than another. On the other hand, the quantum theory of gravity has opened up a new possibility, in which there would be no boundary to space-time and so there would be no need to specify the behavior at the boundary. There would be no singularities at which the laws of science broke down and no edge of space-time at which one would have to appeal to God or some new law to set the boundary conditions for space-time. One could say: "The boundary condition of the universe is that it has no boundary." The universe would be completely self-contained and not affected by anything outside itself. It would neither be created nor destroyed. It would just BE.

It was at the conference in the Vatican mentioned earlier that I first put forward the suggestion that maybe time and space together formed a surface that was finite in size but did not have any boundary or edge. My paper was rather mathematical, however, so its implications for the role of God in the creation of the universe were not generally recognized at the time (just as well for me). At the time of the Vatican conference, I did not know how to use the "no boundary" idea to make predictions about the universe. However I spent the following summer at the University of California, Santa Barbara. There a friend and colleague of mine, Jim Hartle, worked out with me what conditions the universe must satisfy if space-time had no boundary. When I returned to Cambridge, I continued this work with two of my research students, Julian Luttrel and Jonathan Halliwell.

I'd like to emphasize that this idea that time and space should be finite without boundary is just a *proposal:* it cannot be deduced from some other principle. Like any other scientific theory, it may initially be put forward for aesthetic or metaphysical reasons, but the real test is whether it makes predictions that agree with observation. This, however, is difficult to determine in the case of quantum gravity, for two reasons. First, as will be explained in the next chapter, we are not yet sure exactly which theory successfully combines general relativity and quantum mechanics, though we know quite a lot about the form such a theory must have. Second, any model that described the whole universe in detail would be much too complicated mathematically for us to be able to calculate exact predictions. One therefore has to make simplifying assumptions and approximations—and even then, the problem of extracting predictions remains a formidable one.

Each history in the sum over histories will describe not only the space-time but everything in it as well, including any complicated organisms like human beings who can observe the history of the universe. This may provide another justification for the anthropic principle, for if all the histories are possible, then so long as we exist in one of the histories, we may use the anthropic principle to explain why the universe is found to be the way it is. Exactly what meaning can be attached to the other histories, in which we do not exist, is not clear. This view of a quantum theory of gravity would be much more satisfactory, however, if one could show that, using the sum over histories, our universe is not just one of the possible histories but one of the most probable ones. To do this, we must perform the sum over histories for all possible Euclidean space-times that have no boundary.

Under the no boundary proposal one learns that the chance of the universe being found to be following most of the possible histories is negligible, but there is a particular family of histories that are much more probable than the others. These histories may be pictured as being like the surface of the earth, with the distance from the North Pole representing imaginary time and the size of a circle of constant distance from the North Pole representing the spatial size of the universe. The universe starts at the North Pole as a single point. As one moves south, the circles of latitude at constant distance from the North Pole get bigger, corresponding to the universe expanding with imaginary time (Fig. 8.1). The universe would reach a maximum size at the equator and would contract with increasing imaginary time to a single point at the South Pole. Even though the universe would have zero size at the North and South Poles, these points would not be singularities, any more than the North and South Poles on the earth are singular. The laws of science will hold at them, just as they do at the North and South Poles on the earth.

The history of the universe in real time, however, would look very different. At about ten or twenty thousand million years ago, it would have a minimum size, which was equal to

FIGURE 8.1

the maximum radius of the history in imaginary time. At later real times, the universe would expand like the chaotic inflationary model proposed by Linde (but one would not now have to assume that the universe was created somehow in the right sort of state). The universe would expand to a very large size and eventually it would collapse again into what looks like a singularity in real time. Thus, in a sense, we are still all doomed, even if we keep away from black holes. Only if we could picture the universe in terms of imaginary time would there be no singularities.

If the universe really is in such a quantum state, there would be no singularities in the history of the universe in imaginary time. It might seem therefore that my more recent work had completely undone the results of my earlier work on singularities. But, as indicated above, the real importance of the singularity theorems was that they showed that the gravitational field must become so strong that quantum gravitational effects could not be ignored. This in turn led to the idea that the universe could be finite in imaginary time but without boundaries or singularities. When one goes back to the real time in which we live, however, there will still appear to be singularities. The poor astronaut who falls into a black hole will still come to a sticky end; only if he lived in imaginary time would he encounter no singularities.

This might suggest that the so-called imaginary time is really the real time, and that what we call real time is just a figment of our imaginations. In real time, the universe has a beginning and an end at singularities that form a boundary to space-time and at which the laws of science break down. But in imaginary time, there are no singularities or boundaries. So maybe what we call imaginary time is really more basic, and what we call real is just an idea that we invent to help us describe what we think the universe is like. But according to the approach I described in Chapter 1, a scientific theory is just a mathematical model we make to describe our observations: it exists only in our minds. So it is meaningless to ask:

Which is real, "real" or "imaginary" time? It is simply a matter of which is the more useful description.

One can also use the sum over histories, along with the no boundary proposal, to find which properties of the universe are likely to occur together. For example, one can calculate the probability that the universe is expanding at nearly the same rate in all different directions at a time when the density of the universe has its present value. In the simplified models that have been examined so far, this probability turns out to be high; that is, the proposed no boundary condition leads to the prediction that it is extremely probable that the present rate of expansion of the universe is almost the same in each direction. This is consistent with the observations of the microwave background radiation, which show that it has almost exactly the same intensity in any direction. If the universe were expanding faster in some directions than in others, the intensity of the radiation in those directions would be reduced by an additional red shift.

Further predictions of the no boundary condition are currently being worked out. A particularly interesting problem is the size of the small departures from uniform density in the early universe that caused the formation first of the galaxies, then of stars, and finally of us. The uncertainty principle implies that the early universe cannot have been completely uniform because there must have been some uncertainties or fluctuations in the positions and velocities of the particles. Using the no boundary condition, we find that the universe must in fact have started off with just the minimum possible nonuniformity allowed by the uncertainty principle. The universe would have then undergone a period of rapid expansion, as in the inflationary models. During this period, the initial nonuniformities would have been amplified until they were big enough to explain the origin of the structures we observe around us. In an expanding universe in which the density of matter varied slightly from place to place, gravity would have caused the denser regions to slow down their expansion and start contracting. This would lead to the formation of galaxies, stars, and eventually even insignificant creatures like ourselves. Thus all the complicated structures that we see in the universe might be explained by the no boundary condition for the universe together with the uncertainty principle of quantum mechanics.

The idea that space and time may form a closed surface without boundary also has profound implications for the role of God in the affairs of the universe. With the success of scientific theories in describing events, most people have come to believe that God allows the universe to evolve according to a set of laws and does not intervene in the universe to break these laws. However, the laws do not tell us what the universe should have looked like when it started—it would still be up to God to wind up the clockwork and choose how to start it off. So long as the universe had a beginning, we could suppose it had a creator. But if the universe is really completely self-contained, having no boundary or edge, it would have neither beginning nor end: it would simply be. What place, then, for a creator?

9

The Arrow
of Time

In previous chapters we have seen how our views of the
nature of time have changed over the years. Up to the beginning of this century people believed in an absolute time. That
is, each event could be labeled by a number called "time" in
a unique way, and all good clocks would agree on the time
interval between two events. However, the discovery that the
speed of light appeared the same to every observer, no matter
how he was moving, led to the theory of relativity—and in
that one had to abandon the idea that there was a unique
absolute time. Instead, each observer would have his own
measure of time as recorded by a clock that he carried: clocks
carried by different observers would not necessarily agree.
Thus time became a more personal concept, relative to the
observer who measured it.

When one tried to unify gravity with quantum mechanics,
one had to introduce the idea of "imaginary" time. Imaginary time is indistinguishable from directions in space. If one
can go north, one can turn around and head south; equally, if

one can go forward in imaginary time, one ought to be able to turn round and go backward. This means that there can be no important difference between the forward and backward directions of imaginary time. On the other hand, when one looks at "real" time, there's a very big difference between the forward and backward directions, as we all know. Where does this difference between the past and the future come from? Why do we remember the past but not the future?

The laws of science do not distinguish between the past and the future. More precisely, as explained earlier, the laws of science are unchanged under the combination of operations (or symmetries) known as C, P, and T. (C means changing particles for antiparticles. P means taking the mirror image, so left and right are interchanged. And T means reversing the direction of motion of all particles: in effect, running the motion backward.) The laws of science that govern the behavior of matter under all normal situations are unchanged under the combination of the two operations C and P on their own. In other words, life would be just the same for the inhabitants of another planet who were both mirror images of us and who were made of antimatter, rather than matter.

If the laws of science are unchanged by the combination of operations C and P, and also by the combination C, P, and T, they must also be unchanged under the operation T alone. Yet there is a big difference between the forward and backward directions of real time in ordinary life. Imagine a cup of water falling off a table and breaking into pieces on the floor. If you take a film of this, you can easily tell whether it is being run forward or backward. If you run it backward you will see the pieces suddenly gather themselves together off the floor and jump back to form a whole cup on the table. You can tell that the film is being run backward because this kind of behavior is never observed in ordinary life. If it were, crockery manufacturers would go out of business.

The explanation that is usually given as to why we don't see broken cups gathering themselves together off the floor and jumping back onto the table is that it is forbidden by the second law of thermodynamics. This says that in any closed system disorder, or entropy, always increases with time. In other words, it is a form of Murphy's law: Things always tend to go wrong! An intact cup on the table is a state of high order, but a broken cup on the floor is a disordered state. One can go readily from the cup on the table in the past to the broken cup on the floor in the future, but not the other way round.

The increase of disorder or entropy with time is one example of what is called an arrow of time, something that distinguishes the past from the future, giving a direction to time. There are at least three different arrows of time. First, there is the thermodynamic arrow of time, the direction of time in which disorder or entropy increases. Then, there is the psychological arrow of time. This is the direction in which we feel time passes, the direction in which we remember the past but not the future. Finally, there is the cosmological arrow of time. This is the direction of time in which the universe is expanding rather than contracting.

In this chapter I shall argue that the no boundary condition for the universe, together with the weak anthropic principle, can explain why all three arrows point in the same direction—and moreover, why a well-defined arrow of time should exist at all. I shall argue that the psychological arrow is determined by the thermodynamic arrow, and that these two arrows necessarily always point in the same direction. If one assumes the no boundary condition for the universe, we shall see that there must be well-defined thermodynamic and cosmological arrows of time, but they will not point in the same direction for the whole history of the universe. However, I shall argue that it is only when they do point in the same direction that conditions are suitable for the development of intelligent beings who can ask the question: Why does disorder increase in the same direction of time as that in which the universe expands?

I shall discuss first the thermodynamic arrow of time. The second law of thermodynamics results from the fact that there

are always many more disordered states than there are ordered ones. For example, consider the pieces of a jigsaw in a
box. There is one, and only one, arrangement in which the
pieces make a complete picture. On the other hand, there are
a very large number of arrangements in which the pieces are
disordered and don't make a picture.

Suppose a system starts out in one of the small number of
ordered states. As time goes by, the system will evolve
according to the laws of science and its state will change. At
a later time, it is more probable that the system will be in a
disordered state than in an ordered one because there are
more disordered states. Thus disorder will tend to increase
with time if the system obeys an initial condition of high
order.

Suppose the pieces of the jigsaw start off in a box in the
ordered arrangement in which they form a picture. If you
shake the box, the pieces will take up another arrangement.
This will probably be a disordered arrangement in which the
pieces don't form a proper picture, simply because there are
so many more disordered arrangements. Some groups of
pieces may still form parts of the picture, but the more you
shake the box, the more likely it is that these groups will get
broken up and the pieces will be in a completely jumbled
state in which they don't form any sort of picture. So the
disorder of the pieces will probably increase with time if the
pieces obey the initial condition that they start off in a
condition of high order.

Suppose, however, that God decided that the universe
should finish up in a state of high order but that it didn't
matter what state it started in. At early times the universe
would probably be in a disordered state. This would mean
that disorder would *decrease* with time. You would see broken cups gathering themselves together and jumping back
onto the table. However, any human beings who were observing the cups would be living in a universe in which
disorder decreased with time. I shall argue that such beings
would have a psychological arrow of time that was backward.

That is, they would remember events in the future, and not remember events in their past. When the cup was broken, they would remember it being on the table, but when it was on the table, they would not remember it being on the floor.

It is rather difficult to talk about human memory because we don't know how the brain works in detail. We do, however, know all about how computer memories work. I shall therefore discuss the psychological arrow of time for computers. I think it is reasonable to assume that the arrow for computers is the same as that for humans. If it were not, one could make a killing on the stock exchange by having a computer that would remember tomorrow's prices!

A computer memory is basically a device containing elements that can exist in either of two states. A simple example is an abacus. In its simplest form, this consists of a number of wires; on each wire is a bead that can be put in one of two positions. Before an item is recorded in a computer's memory, the memory is in a disordered state, with equal probabilities for the two possible states. (The abacus beads are scattered randomly on the wires of the abacus.) After the memory interacts with the system to be remembered, it will definitely be in one state or the other, according to the state of the system. (Each abacus bead will be at either the left or the right of the abacus wire.) So the memory has passed from a disordered state to an ordered one. However, in order to make sure that the memory is in the right state, it is necessary to use a certain amount of energy (to move the bead or to power the computer, for example). This energy is dissipated as heat, and increases the amount of disorder in the universe. One can show that this increase in disorder is always greater than the increase in the order of the memory itself. Thus the heat expelled by the computer's cooling fan means that when a computer records an item in memory, the total amount of disorder in the universe still goes up. The direction of time in which a computer remembers the past is the same as that in which disorder increases.

Our subjective sense of the direction of time, the psychological arrow of time, is therefore determined within our brain by the thermodynamic arrow of time. Just as a computer, we must remember things in the order in which entropy increases. This makes the second law of thermodynamics almost trivial. Disorder increases with time because we measure time in the direction in which disorder increases. You can't have a safer bet than that!

But why should the thermodynamic arrow of time exist at all? Or, in other words, why should the universe be in a state of high order at one end of time, the end that we call the past? Why is it not in a state of complete disorder at all times? After all, this might seem more probable. And why is the direction of time in which disorder increases the same as that in which the universe expands?

In the classical theory of general relativity one cannot predict how the universe would have begun because all the known laws of science would have broken down at the big bang singularity. The universe could have started out in a very smooth and ordered state. This would have led to well-defined thermodynamic and cosmological arrows of time, as we observe. But it could equally well have started out in a very lumpy and disordered state. In that case, the universe would already be in a state of complete disorder, so disorder could not increase with time. It would either stay constant, in which case there would be no well-defined thermodynamic arrow of time, or it would decrease, in which case the thermodynamic arrow of time would point in the opposite direction to the cosmological arrow. Neither of these possibilities agrees with what we observe. However, as we have seen, classical general relativity predicts its own downfall. When the curvature of space-time becomes large, quantum gravitational effects will become important and the classical theory will cease to be a good description of the universe. One has to use a quantum theory of gravity to understand how the universe began.

In a quantum theory of gravity, as we saw in the last chapter, in order to specify the state of the universe one would still have to say how the possible histories of the universe would behave at the boundary of space-time in the past. One could avoid this difficulty of having to describe what we do not and cannot know only if the histories satisfy the no boundary condition: they are finite in extent but have no boundaries, edges, or singularities. In that case, the beginning of time would be a regular, smooth point of space-time and the universe would have begun its expansion in a very smooth and ordered state. It could not have been completely uniform, because that would violate the uncertainty principle of quantum theory. There had to be small fluctuations in the density and velocities of particles. The no boundary condition, however, implied that these fluctuations were as small as they could be, consistent with the uncertainty principle.

The universe would have started off with a period of exponential or ''inflationary'' expansion in which it would have increased its size by a very large factor. During this expansion, the density fluctuations would have remained small at first, but later would have started to grow. Regions in which the density was slightly higher than average would have had their expansion slowed down by the gravitational attraction of the extra mass. Eventually, such regions would stop expanding and collapse to form galaxies, stars, and beings like us. The universe would have started in a smooth and ordered state, and would become lumpy and disordered as time went on. This would explain the existence of the thermodynamic arrow of time.

But what would happen if and when the universe stopped expanding and began to contract? Would the thermodynamic arrow reverse and disorder begin to decrease with time? This would lead to all sorts of science-fictionlike possibilities for people who survived from the expanding to the contracting phase. Would they see broken cups gathering themselves together off the floor and jumping back onto the table? Would they be able to remember tomorrow's prices and make a fortune on the stock market? It might seem a bit academic to worry about what would happen when the universe collapses again, as it will not start to contract for at least another ten thousand million years. But there is a quicker way to find out what will happen: jump into a black hole. The collapse of a star to form a black hole is rather like the later stages of the collapse of the whole universe. So if disorder were to decrease in the contracting phase of the universe, one might also expect it to decrease inside a black hole. So perhaps an astronaut who fell into a black hole would be able to make money at roulette by remembering where the ball went before he placed his bet. (Unfortunately, however, he would not have long to play before he was turned to spaghetti. Nor would he be able to let us know about the reversal of the thermodynamic arrow, or even bank his winnings, because he would be trapped behind the event horizon of the black hole.)

At first, I believed that disorder would decrease when the universe recollapsed. This was because I thought that the universe had to return to a smooth and ordered state when it became small again. This would mean that the contracting phase would be like the time reverse of the expanding phase. People in the contracting phase would live their lives backward: they would die before they were born and get younger as the universe contracted.

This idea is attractive because it would mean a nice symmetry between the expanding and contracting phases. However, one cannot adopt it on its own, independent of other ideas about the universe. The question is: Is it implied by the no boundary condition, or is it inconsistent with that condition? As I said, I thought at first that the no boundary condition did indeed imply that disorder would decrease in the contracting phase. I was misled partly by the analogy with the surface of the earth. If one took the beginning of the universe to correspond to the North Pole, then the end of the universe should be similar to the beginning, just as the South Pole is similar to the North. However, the North and South Poles correspond to the beginning and end of the universe in imaginary time. The beginning and end in real time can be very different from each other. I was also misled by work I had done on a simple model of the universe in which the collapsing phase looked like the time reverse of the expanding phase. However, a colleague of mine, Don Page, of Penn State University, pointed out that the no boundary condition did not require the contracting phase necessarily to be the time reverse of the expanding phase. Further, one of my students, Raymond Laflamme, found that in a slightly more complicated model, the collapse of the universe was very different from the expansion. I realized that I had made a mistake: the no boundary condition implied that disorder would in fact continue to increase during the contraction. The thermodynamic and psychological arrows of time would not reverse when the universe begins to recontract or inside black holes.

What should you do when you find you have made a mistake like that? Some people never admit that they are wrong and continue to find new, and often mutually inconsistent, arguments to support their case—as Eddington did in opposing black hole theory. Others claim to have never really supported the incorrect view in the first place or, if they did, it was only to show that it was inconsistent. It seems to me much better and less confusing if you admit in print that you were wrong. A good example of this was Einstein, who called the cosmological constant, which he introduced when he was trying to make a static model of the universe, the biggest mistake of his life.

To return to the arrow of time, there remains the question: Why do we observe that the thermodynamic and cosmological arrows point in the same direction? Or in other words, why does disorder increase in the same direction of time as that in which the universe expands? If one believes that the universe will expand and then contract again, as the no boundary proposal seems to imply, this becomes a question of why we should be in the expanding phase rather than the contracting phase.

One can answer this on the basis of the weak anthropic principle. Conditions in the contracting phase would not be suitable for the existence of intelligent beings who could ask the question: Why is disorder increasing in the same direction of time as that in which the universe is expanding? The inflation in the early stages of the universe, which the no boundary proposal predicts, means that the universe must be expanding at very close to the critical rate at which it would just avoid recollapse, and so will not recollapse for a very long time. By then all the stars will have burned out and the protons and neutrons in them will probably have decayed into light particles and radiation. The universe would be in a state of almost complete disorder. There would be no strong thermodynamic arrow of time. Disorder couldn't increase much because the universe would be in a state of almost complete disorder already. However, a strong thermodynamic arrow is necessary for intelligent life to operate. In order to survive, human beings have to consume food, which is an ordered form of energy, and convert it into heat, which is a disordered form of energy. Thus intelligent life could not exist in the contracting phase of the universe. This is the explanation of why we observe that the thermodynamic and cosmological arrows of time point in the same direction. It is not that the expansion of the universe causes disorder to increase. Rather, it is that the no boundary condition causes disorder to increase and the conditions to be suitable for intelligent life only in the expanding phase.

To summarize, the laws of science do not distinguish between the forward and backward directions of time. However, there are at least three arrows of time that do distinguish the past from the future. They are the thermodynamic arrow, the direction of time in which disorder increases; the psychological arrow, the direction of time in which we remember the past and not the future; and the cosmological arrow, the direction of time in which the universe expands rather than contracts. I have shown that the psychological arrow is essentially the same as the thermodynamic arrow, so that the two would always point in the same direction. The no boundary proposal for the universe predicts the existence of a well-defined thermodynamic arrow of time because the universe must start off in a smooth and ordered state. And the reason we observe this thermodynamic arrow to agree with the cosmological arrow is that intelligent beings can exist only in the expanding phase. The contracting phase will be unsuitable because it has no strong thermodynamic arrow of time.

The progress of the human race in understanding the universe has established a small corner of order in an increasingly disordered universe. If you remember every word in this book, your memory will have recorded about two million pieces of information: the order in your brain will have increased by about two million units. However, while you have been reading the book, you will have converted at least a thousand calories of ordered energy, in the form of food, into disordered energy, in the form of heat that you lose to the air around you by convection and sweat. This will increase the disorder of the universe by about twenty million million million million units—or about ten million million million times the increase in order in your brain—and that's if you remember *everything* in this book. In the next chapter I will try to increase the order in our neck of the woods a little further by explaining how people are trying to fit together the partial theories I have described to form a complete unified theory that would cover everything in the universe.

10

The Unification
of Physics

As was explained in the first chapter, it would be very difficult to construct a complete unified theory of everything in the universe all at one go. So instead we have made progress by finding partial theories that describe a limited range of happenings and by neglecting other effects or approximating them by certain numbers. (Chemistry, for example, allows us to calculate the interactions of atoms, without knowing the internal structure of an atom's nucleus.) Ultimately however, one would hope to find a complete, consistent, unified theory that would include all these partial theories as approximations, and that did not need to be adjusted to fit the facts by picking the values of certain arbitrary numbers in the theory. The quest for such a theory is known as "the unification of physics." Einstein spent most of his later years unsuccessfully searching for a unified theory, but the time was not ripe: there were partial theories for gravity and the electromagnetic force, but very little was known about the nuclear forces. Moreover, Einstein refused to believe in the reality of quantum mechanics, despite the important role he had played in its development. Yet it seems that the uncertainty principle is a fundamental feature of the universe we live in. A successful unified theory must therefore neccessarily incorporate this principle.

As I shall describe, the prospects for finding such a theory seem to be much better now because we know so much more about the universe. But we must beware of overconfidence—we have had false dawns before! At the beginning of this century, for example, it was thought that everything could be explained in terms of the properties of continuous matter, such as elasticity and heat conduction. The discovery of atomic structure and the uncertainty principle put an emphatic end to that. Then again, in 1928, physicist and Nobel prize winner Max Born told a group of visitors to Göttingen University, "Physics, as we know it, will be over in six months." His confidence was based on the recent discovery by Dirac of the equation that governed the electron. It was thought that a similar equation would govern the proton, which was the only other particle known at the time, and that would be the end of theoretical physics. However, the discovery of the neutron and of nuclear forces knocked that one on the head, too. Having said this, I still believe there are grounds for cautious optimism that we may now be near the end of the search for the ultimate laws of nature.

In previous chapters I have described general relativity, the partial theory of gravity, and the partial theories that govern the weak, the strong, and the electromagnetic forces. The last three may be combined in so-called grand unified theories, or GUTs, which are not very satisfactory because they do not include gravity and because they contain a number of quantities, like the relative masses of different particles, that cannot be predicted from the theory but have to be chosen to fit observations. The main difficulty in finding a theory that unifies gravity with the other forces is that general relativity is a "classical" theory; that is, it does not incorporate the uncertainty principle of quantum mechanics. On the other hand, the other partial theories depend on quantum mechanics in an essential way. A necessary first step, therefore, is to combine general relativity with the uncertainty principle. As we have seen, this can produce some remarkable consequences, such as black holes not being black, and the universe not having any singularities but being completely self-contained and without a boundary. The trouble is, as explained in Chapter 7, the uncertainty principle means that even "empty" space is filled with pairs of virtual particles and antiparticles. These pairs would have an infinite amount of energy and, therefore, by Einstein's famous equation $E = mc^2$, they would have an infinite amount of mass. Their gravitational attraction would thus curve up the universe to infinitely small size.

Rather similar, seemingly absurd infinities occur in the other partial theories, but in all these cases the infinities can be cancelled out by a process called renormalization. This involves cancelling the infinities by introducing other infinities. Although this technique is rather dubious mathematically, it does seem to work in practice, and has been used with these theories to make predictions that agree with observations to an extraordinary degree of accuracy. Renormalization, however, does have a serious drawback from the point of view of trying to find a complete theory, because it means that the actual values of the masses and the strengths of the forces cannot be predicted from the theory, but have to be chosen to fit the observations.

In attempting to incorporate the uncertainty principle into general relativity, one has only two quantities that can be adjusted: the strength of gravity and the value of the cosmological constant. But adjusting these is not sufficient to remove all the infinities. One therefore has a theory that seems to predict that certain quantities, such as the curvature of space-time, are really infinite, yet these quantities can be observed and measured to be perfectly finite! This problem in combining general relativity and the uncertainty principle had been suspected for some time, but was finally confirmed by detailed calculations in 1972. Four years later, a possible solution, called "supergravity," was suggested. The idea was to combine the spin-2 particle called the graviton, which carries the gravitational force, with certain other new particles of spin 3/2, 1, 1/2, and 0. In a sense, all these particles could then be regarded as different aspects of the same "superparticle," thus unifying the matter particles with spin 1/2 and 3/2 with the force-carrying particles of spin 0, 1, and 2. The virtual particle/antiparticle pairs of spin 1/2 and 3/2 would have negative energy, and so would tend to cancel out the positive energy of the spin 2, 1, and 0 virtual pairs. This would cause many of the possible infinities to cancel out, but it was

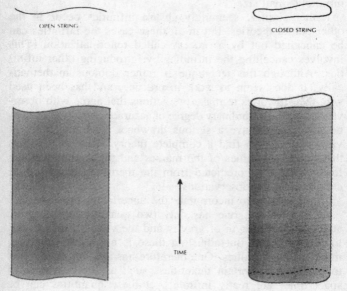

OPEN STRING

CLOSED STRING

WORLD-SHEET OF OPEN STRING

TIME

WORLD-SHEET OF CLOSED STRING

FIGURE 10.1 AND FIGURE 10.2

suspected that some infinities might still remain. However, the calculations required to find out whether or not there were any infinities left uncanceled were so long and difficult that no one was prepared to undertake them. Even with a computer, it was reckoned it would take at least four years, and the chances were very high that one would make at least one mistake, probably more. So one would know one had the right answer only if someone else repeated the calculation and got the same answer, and that did not seem very likely!

Despite these problems, and the fact that the particles in the supergravity theories did not seem to match the observed

ONE SINGLE STRING

THE TWO STRINGS JOIN

TWO SEPARATE
STRINGS

TIME

WORLD-SHEET OF TWO OPEN STRINGS JOINING

FIGURE 10.3

particles, most scientists believed that supergravity was probably the right answer to the problem of the unification of physics. It seemed the best way of unifying gravity with the other forces. However, in 1984 there was a remarkable change of opinion in favor of what are called string theories. In these theories the basic objects are not particles, which occupy a single point of space, but things that have a length but no other dimension, like an infinitely thin piece of string. These strings may have ends (the so-called open strings) or they may be joined up with themselves in closed loops (closed strings) (Fig. 10.1 and Fig. 10.2). A particle occupies one point of space at each instant of time. Thus its history can be represented by a line in space-time (the "world-line"). A string, on the other hand, occupies a line in space at each moment of time. So its history in space-time is a two-dimensional surface called the world-sheet. (Any point on such a world-sheet can be described by two numbers: one specifying the time and the other the position of the point on the string.) The world-sheet of an open string is a strip; its edges represent the paths through space-time of the ends of the string (Fig. 10.1). The world-sheet of a closed string is a cylinder or tube (Fig. 10.2); a slice through the tube is a circle, which represents the position of the string at one particular time.

Two pieces of string can join together to form a single string; in the case of open strings they simply join at the ends (Fig. 10.3), while in the case of closed strings it is like the two legs joining on a pair of trousers (Fig. 10.4). Similarly, a single piece of string can divide into two strings. In string theories, what were previously thought of as particles are now pictured as waves traveling down the string, like waves on a vibrating kite string. The emission or absorption of one particle by another corresponds to the dividing or joining together of strings. For example, the gravitational force of the sun on the earth was pictured in particle theories as being caused by the emission of a graviton by a particle in the sun and its absorption by a particle in the earth (Fig. 10.5). In string

ONE SINGLE STRING

THE TWO STRINGS JOIN

TWO SEPARATE STRINGS

TIME

WORLD-SHEET OF TWO CLOSED STRINGS JOINING

FIGURE 10.4

theory, this process corresponds to an H-shaped tube or pipe (Fig. 10.6) (string theory is rather like plumbing, in a way). The two vertical sides of the H correspond to the particles in the sun and the earth and the horizontal crossbar corresponds to the graviton that travels between them.

String theory has a curious history. It was originally invented in the late 1960s in an attempt to find a theory to describe the strong force. The idea was that particles like the proton and the neutron could be regarded as waves on a string. The strong forces between the particles would correspond to pieces of string that went between other bits of string, as in a spider's web. For this theory to give the

FIGURE 10.5 AND FIGURE 10.6

observed value of the strong force between particles, the strings had to be like rubber bands with a pull of about ten tons.

In 1974 Joël Scherk from Paris and John Schwarz from the California Institute of Technology published a paper in which they showed that string theory could describe the gravitational force, but only if the tension in the string were very much higher, about a thousand million million million million million million tons (1 with thirty-nine zeros after it). The predictions of the string theory would be just the same as those of general relativity, on normal length scales, but they would differ at very small distances, less than a thousand million million million million millionth of a centimeter (a centimeter divided by 1 with thirty-three zeros after it). Their work did not receive much attention, however, because at just about that time most people abandoned the original string theory of the strong force in favor of the theory based on quarks and gluons, which seemed to fit much better with observations. Scherk died in tragic circumstances (he suffered from diabetes and went into a coma when no one was around to give him an injection of insulin). So Schwarz was left alone as almost the only supporter of string theory, but now with the much higher proposed value of the string tension.

In 1984 interest in strings suddenly revived, apparently for two reasons. One was that people were not really making much progress toward showing that supergravity was finite or that it could explain the kinds of particles that we observe. The other was the publication of a paper by John Schwarz and Mike Green of Queen Mary College, London, that showed that string theory might be able to explain the existence of particles that have a built-in left-handedness, like some of the particles that we observe. Whatever the reasons, a large number of people soon began to work on string theory and a new version was developed, the so-called heterotic string, which seemed as if it might be able to explain the types of particles that we observe.

String theories also lead to infinities, but it is thought they

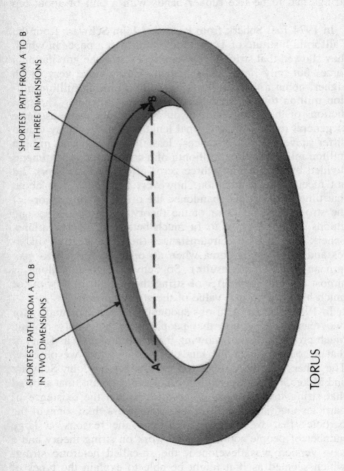

SHORTEST PATH FROM A TO B
IN THREE DIMENSIONS

SHORTEST PATH FROM A TO B
IN TWO DIMENSIONS

TORUS

FIGURE 10.7

will all cancel out in versions like the heterotic string (though this is not yet known for certain). String theories, however, have a bigger problem: they seem to be consistent only if space-time has either ten or twenty-six dimensions, instead of the usual four! Of course, extra space-time dimensions are a commonplace of science fiction; indeed, they are almost a necessity, since otherwise the fact that relativity implies that one cannot travel faster than light means that it would take far too long to travel between stars and galaxies. The science fiction idea is that maybe one can take a shortcut through a higher dimension. One can picture this in the following way. Imagine that the space we live in has only two dimensions and is curved like the surface of an anchor ring or torus (Fig. 10.7). If you were on one side of the inside edge of the ring and you wanted to get to a point on the other side, you would have to go round the inner edge of the ring. However, if you were able to travel in the third dimension, you could cut straight across.

Why don't we notice all these extra dimensions, if they are really there? Why do we see only three space and one time dimension? The suggestion is that the other dimensions are curved up into a space of very small size, something like a million million million million millionth of an inch. This is so small that we just don't notice it; we see only one time and three space dimensions, in which space-time is fairly flat. It is like the surface of an orange: if you look at it close up, it is all curved and wrinkled, but if you look at it from a distance, you don't see the bumps and it appears to be smooth. So it is with space-time: on a very small scale it is ten-dimensional and highly curved, but on bigger scales you don't see the curvature or the extra dimensions. If this picture is correct, it spells bad news for would-be space travelers: the extra dimensions would be far too small to allow a spaceship through. However, it raises another major problem. Why should some, but not all, of the dimensions be curled up into a small ball? Presumably, in the very early universe all the dimensions would have been very curved. Why did one time and three

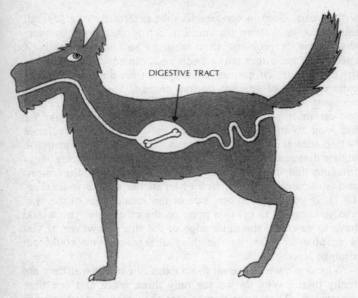

DIGESTIVE TRACT

TWO-DIMENSIONAL ANIMAL

FIGURE 10.8

space dimensions flatten out, while the other dimensions remain tightly curled up?

One possible answer is the anthropic principle. Two space dimensions do not seem to be enough to allow for the development of complicated beings like us. For example, two-dimensional animals living on a one-dimensional earth would have to climb over each other in order to get past each other. If a two-dimensional creature ate something it could not digest completely, it would have to bring up the remains the same way it swallowed them, because if there were a passage right through its body, it would divide the creature into two separate halves; our two-dimensional being would fall apart (Fig. 10.8). Similarly, it is difficult to see how there could be any circulation of the blood in a two-dimensional creature.

There would also be problems with more than three space dimensions. The gravitational force between two bodies would decrease more rapidly with distance than it does in three dimensions. (In three dimensions, the gravitational force drops to $\frac{1}{4}$ if one doubles the distance. In four dimensions it would drop to $\frac{1}{8}$, in five dimensions to $\frac{1}{16}$, and so on.) The significance of this is that the orbits of planets, like the earth, around the sun would be unstable: the least disturbance from a circular orbit (such as would be caused by the gravitational attraction of other planets) would result in the earth spiraling away from or into the sun. We would either freeze or be burned up. In fact, the same behavior of gravity with distance in more than three space dimensions means that the sun would not be able to exist in a stable state with pressure balancing gravity. It would either fall apart or it would collapse to form a black hole. In either case, it would not be of much use as a source of heat and light for life on Earth. On a smaller scale, the electrical forces that cause the electrons to orbit round the nucleus in an atom would behave in the same way as gravitational forces. Thus the electrons would either escape from the atom altogether or would spiral into the nucleus. In either case, one could not have atoms as we know them.

It seems clear then that life, at least as we know it, can exist only in regions of space-time in which one time and three space dimensions are not curled up small. This would mean that one could appeal to the weak anthropic principle, provided one could show that string theory does at least allow there to be such regions of the universe—and it seems that indeed string theory does. There may well be other regions of the universe, or other universes (whatever *that* may mean), in which all the dimensions are curled up small or in which more than four dimensions are nearly flat, but there would be no intelligent beings in such regions to observe the different number of effective dimensions.

Apart from the question of the number of dimensions that space-time appears to have, string theory still has several other problems that must be solved before it can be acclaimed as the ultimate unified theory of physics. We do not yet know whether all the infinities do cancel each other out, or exactly how to relate the waves on the string to the particular types of particle that we observe. Nevertheless, it is likely that answers to these questions will be found over the next few years, and that by the end of the century we shall know whether string theory is indeed the long sought-after unified theory of physics.

But can there really be such a unified theory? Or are we perhaps just chasing a mirage? There seems to be three possibilities:

1) There really is a complete unified theory, which we will someday discover if we are smart enough.

2) There is no ultimate theory of the universe, just an infinite sequence of theories that describe the universe more and more accurately.

3) There is no theory of the universe; events cannot be predicted beyond a certain extent but occur in a random and arbitrary manner.

Some would argue for the third possibility on the grounds that if there were a complete set of laws, that would infringe God's freedom to change his mind and intervene in the world. It's a bit like the old paradox: Can God make a stone so heavy that he can't lift it? But the idea that God might want to change his mind is an example of the fallacy, pointed out by St. Augustine, of imagining God as a being existing in time: time is a property only of the universe that God created. Presumably, he knew what he intended when he set it up!

With the advent of quantum mechanics, we have come to recognize that events cannot be predicted with complete accuracy but that there is always a degree of uncertainty. If one likes, one could ascribe this randomness to the intervention of God, but it would be a very strange kind of intervention: there is no evidence that it is directed toward any purpose. Indeed, if it were, it would by definition not be random. In modern times, we have effectively removed the third possibility above by redefining the goal of science: our aim is to formulate a set of laws that enables us to predict events only up to the limit set by the uncertainty principle.

The second possibility, that there is an infinite sequence of more and more refined theories, is in agreement with all our experience so far. On many occasions we have increased the sensitivity of our measurements or made a new class of observations, only to discover new phenomena that were not predicted by the existing theory, and to account for these we have had to develop a more advanced theory. It would therefore not be very surprising if the present generation of grand unified theories was wrong in claiming that nothing essentially new will happen between the electroweak unification energy of about 100 GeV and the grand unification energy of about a thousand million million GeV. We might indeed expect to find several new layers of structure more basic than the quarks and electrons that we now regard as "elementary" particles.

However, it seems that gravity may provide a limit to this sequence of "boxes within boxes." If one had a particle with an energy above what is called the Planck energy, ten million million million GeV (1 followed by nineteen zeros), its mass would be so concentrated that it would cut itself off from the rest of the universe and form a little black hole. Thus it does seem that the sequence of more and more refined theories should have some limit as we go to higher and higher energies, so that there should be some ultimate theory of the universe. Of course, the Planck energy is a very long way from the energies of around a hundred GeV, which are the most that we can produce in the laboratory at the present time. We shall not bridge that gap with particle accelerators in the foreseeable future! The very early stages of the universe, however, are an arena where such energies must have occurred. I think that there is a good chance that the study of the early universe and the requirements of mathematical consistency will lead us to a complete unified theory within the lifetime of some of us who are around today, always presuming we don't blow ourselves up first.

What would it mean if we actually did discover the ultimate theory of the universe? As was explained in Chapter 1, we could never be quite sure that we had indeed found the correct theory, since theories can't be proved. But if the theory was mathematically consistent and always gave predictions that agreed with observations, we could be reasonably confident that it was the right one. It would bring to an end a long and glorious chapter in the history of humanity's intellectual struggle to understand the universe. But it would also revolutionize the ordinary person's understanding of the laws that govern the universe. In Newton's time it was possible for an educated person to have a grasp of the whole of human knowledge, at least in outline. But since then, the pace of the development of science has made this impossible. Because theories are always being changed to account for new observations, they are never properly digested or simplified so that ordinary people can understand them. You have to be a specialist, and even then, you can only hope to have a proper grasp of a small proportion of the scientific theories. Further, the rate of progress is so rapid that what one learns at school or university is always a bit out of date. Only a few people can keep up with the rapidly advancing frontier of knowledge, and they have to devote their whole time to it and specialize in a small area. The rest of the population has little idea of the advances that are being made or the excitement they are generating. Seventy years ago, if Eddington is to be believed, only two people understood the general theory of relativity. Nowadays tens of thousands of university graduates do, and many millions of people are at least familiar with the idea. If a complete unified theory was discovered, it would only be a matter of time before it was digested and simplified in the same way and taught in schools, at least in outline. We would then all be able to have some understanding of the laws that govern the universe and are responsible for our existence.

Even if we do discover a complete unified theory, it would not mean that we would be able to predict events in general, for two reasons. The first is the limitation that the uncertainty principle of quantum mechanics sets on our powers of prediction. There is nothing we can do to get around that. In practice, however, this first limitation is less restrictive than the second one. It arises from the fact that we could not solve the equations of the theory exactly, except in very simple situations. (We cannot even solve exactly for the motion of three bodies in Newton's theory of gravity, and the difficulty increases with the number of bodies and the complexity of the theory.) We already know the laws that govern the behavior of matter under all but the most extreme conditions. In particular, we know the basic laws that underlie all of chemistry and biology. Yet we have certainly not reduced these subjects to the status of solved problems; we have, as yet, had little success in predicting human behavior from mathematical equations! So even if we do find a complete set of basic laws, there will still be in the years ahead the intellectually challenging task of developing better approximation methods, so that we can make useful predictions of the probable outcomes in complicated and realistic situations. A complete, consistent, unified theory is only the first step: our goal is a complete *understanding* of the events around us, and of our own existence.

11

Conclusion

We find ourselves in a bewildering world. We want to make sense of what we see around us and to ask: What is the nature of the universe? What is our place in it and where did it and we come from? Why is it the way it is?

To try to answer these questions we adopt some "world picture." Just as an infinite tower of tortoises supporting the flat earth is such a picture, so is the theory of superstrings. Both are theories of the universe, though the latter is much more mathematical and precise than the former. Both theories lack observational evidence: no one has ever seen a giant tortoise with the earth on its back, but then, no one has seen a superstring either. However, the tortoise theory fails to be a good scientific theory because it predicts that people should be able to fall off the edge of the world. This has not been found to agree with experience, unless that turns out to be the explanation for the people who are supposed to have disappeared in the Bermuda Triangle!

The earliest theoretical attempts to describe and explain the universe involved the idea that events and natural phenomena were controlled by spirits with human emotions who acted in a very humanlike and unpredictable manner. These spirits inhabited natural objects, like rivers and mountains, including celestial bodies, like the sun and moon. They had to be placated and their favors sought in order to ensure the fertility of the soil and the rotation of the seasons. Gradually, however, it must have been noticed that there were certain regularities: the sun always rose in the east and set in the west, whether or not a sacrifice had been made to the sun god. Further, the sun, the moon, and the planets followed precise paths across the sky that could be predicted in advance with considerable accuracy. The sun and the moon might still be gods, but they were gods who obeyed strict laws, apparently without any exceptions, if one discounts stories like that of the sun stopping for Joshua.

At first, these regularities and laws were obvious only in astronomy and a few other situations. However, as civilization developed, and particularly in the last 300 years, more and more regularities and laws were discovered. The success of these laws led Laplace at the beginning of the nineteenth century to postulate scientific determinism, that is, he suggested that there would be a set of laws that would determine the evolution of the universe precisely, given its configuration at one time.

Laplace's determinism was incomplete in two ways. It did not say how the laws should be chosen and it did not specify the initial configuration of the universe. These were left to God. God would choose how the universe began and what laws it obeyed, but he would not intervene in the universe once it had started. In effect, God was confined to the areas that nineteenth-century science did not understand.

We now know that Laplace's hopes of determinism cannot be realized, at least in the terms he had in mind. The uncertainty principle of quantum mechanics implies that certain pairs of quantities, such as the position and velocity of a particle, cannot both be predicted with complete accuracy.

Quantum mechanics deals with this situation via a class of quantum theories in which particles don't have well-defined positions and velocities but are represented by a wave. These quantum theories are deterministic in the sense that they give laws for the evolution of the wave with time. Thus if one knows the wave at one time, one can calculate it at any other time. The unpredictable, random element comes in only when we try to interpret the wave in terms of the positions and velocities of particles. But maybe that is our mistake: maybe there are no particle positions and velocities, but only waves. It is just that we try to fit the waves to our preconceived ideas of positions and velocities. The resulting mismatch is the cause of the apparent unpredictability.

In effect, we have redefined the task of science to be the discovery of laws that will enable us to predict events up to the limits set by the uncertainty principle. The question remains, however: How or why were the laws and the initial state of the universe chosen?

In this book I have given special prominence to the laws that govern gravity, because it is gravity that shapes the large-scale structure of the universe, even though it is the weakest of the four categories of forces. The laws of gravity were incompatible with the view held until quite recently that the universe is unchanging in time: the fact that gravity is always attractive implies that the universe must be either expanding or contracting. According to the general theory of relativity, there must have been a state of infinite density in the past, the big bang, which would have been an effective beginning of time. Similarly, if the whole universe recollapsed, there must be another state of infinite density in the future, the big crunch, which would be an end of time. Even if the whole universe did not recollapse, there would be singularities in any localized regions that collapsed to form black holes. These singularities would be an end of time for anyone who fell into the black hole. At the big bang and other singularities, all the laws would have broken down, so God would still have had complete freedom to choose what happened and how the universe began.

When we combine quantum mechanics with general relativity, there seems to be a new possibility that did not arise before: that space and time together might form a finite, four-dimensional space without singularities or boundaries, like the surface of the earth but with more dimensions. It seems that this idea could explain many of the observed features of the universe, such as its large-scale uniformity and also the smaller-scale departures from homogeneity, like galaxies, stars, and even human beings. It could even account for the arrow of time that we observe. But if the universe is completely self-contained, with no singularities or boundaries, and completely described by a unified theory, that has profound implications for the role of God as Creator.

Einstein once asked the question: "How much choice did God have in constructing the universe?" If the no boundary proposal is correct, he had no freedom at all to choose initial conditions. He would, of course, still have had the freedom to choose the laws that the universe obeyed. This, however, may not really have been all that much of a choice; there may well be only one, or a small number, of complete unified theories, such as the heterotic string theory, that are self-consistent and allow the existence of structures as complicated as human beings who can investigate the laws of the universe and ask about the nature of God.

Even if there is only one possible unified theory, it is just a set of rules and equations. What is it that breathes fire into the equations and makes a universe for them to describe? The usual approach of science of constructing a mathematical model cannot answer the questions of why there should be a universe for the model to describe. Why does the universe go to all the bother of existing? Is the unified theory so compelling that it brings about its own existence? Or does it need a creator, and, if so, does he have any other effect on the universe? And who created him?

Up to now, most scientists have been too occupied with the

development of new theories that describe *what* the universe is to ask the question *why*. On the other hand, the people whose business it is to ask *why*, the philosophers, have not been able to keep up with the advance of scientific theories. In the eighteenth century, philosophers considered the whole of human knowledge, including science, to be their field and discussed questions such as: Did the universe have a beginning? However, in the nineteenth and twentieth centuries, science became too technical and mathematical for the philosophers, or anyone else except a few specialists. Philosophers reduced the scope of their inquiries so much that Wittgenstein, the most famous philosopher of this century, said "The sole remaining task for philosophy is the analysis of language." What a comedown from the great tradition of philosophy from Aristotle to Kant!

However, if we do discover a complete theory, it should in time be understandable in broad principle by everyone, not just a few scientists. Then we shall all, philosophers, scientists, and just ordinary people, be able to take part in the discussion of the question of why it is that we and the universe exist. If we find the answer to that, it would be the ultimate triumph of human reason—for then we would know the mind of God.

Albert Einstein

Einstein's connection with the politics of the nuclear bomb is well known: he signed the famous letter to President Franklin Roosevelt that persuaded the United States to take the idea seriously, and he engaged in postwar efforts to prevent nuclear war. But these were not just the isolated actions of a scientist dragged into the world of politics. Einstein's life was, in fact, to use his own words, "divided between politics and equations."

Einstein's earliest political activity came during the First World War, when he was a professor in Berlin. Sickened by what he saw as the waste of human lives, he became involved in antiwar demonstrations. His advocacy of civil disobedience and public encouragement of people to refuse conscription did little to endear him to his colleagues. Then, following the war, he directed his efforts toward reconciliation and improving international relations. This, too, did not make him popular, and soon his politics were making it difficult for him to visit the United States, even to give lectures.

Einstein's second great cause was Zionism. Although he was Jewish by descent, Einstein rejected the biblical idea of God. However, a growing awareness of anti-Semitism, both before and during the First World War, led him gradually to identify with the Jewish community, and later to become an outspoken supporter of Zionism. Once more unpopularity did not stop him from speaking his mind. His theories came under attack; an anti-Einstein organization was even set up. One man was convicted of inciting others to murder Einstein (and fined a mere six dollars). But Einstein was phlegmatic: when a book was published entitled *100 Authors Against Einstein*, he retorted, "If I were wrong, then one would have been enough!"

In 1933, Hitler came to power. Einstein was in America, and declared he would not return to Germany. Then, while Nazi militia raided his house and confiscated his bank account, a Berlin newspaper displayed the headline "Good News from Einstein—He's Not Coming Back." In the face of the Nazi threat, Einstein renounced pacifism, and eventually, fearing that German scientists would build a nuclear bomb, proposed that the United States should develop its own. But even before the first atomic bomb had been detonated, he was publicly warning of the dangers of nuclear war and proposing international control of nuclear weaponry.

Throughout his life, Einstein's efforts toward peace probably achieved little that would last—and certainly won him few friends. His vocal support of the Zionist cause, however, was duly recognized in 1952, when he was offered the presidency of Israel. He declined, saying he thought he was too naive in politics. But perhaps his real reason was different: to quote him again, "Equations are more important to me, because politics is for the present, but an equation is something for eternity."

GALILEO GALILEI

Galileo Galilei

Galileo, perhaps more than any other single person, was responsible for the birth of modern science. His renowned conflict with the Catholic Church was central to his philosophy, for Galileo was one of the first to argue that man could hope to understand how the world works, and, moreover, that we could do this by observing the real world.

Galileo had believed Copernican theory (that the planets orbited the sun) since early on, but it was only when he found the evidence needed to support the idea that he started to publicly support it. He wrote about Copernicus's theory in Italian (not the usual academic Latin), and soon his views became widely supported outside the universities. This annoyed the Aristotelian professors, who united against him seeking to persuade the Catholic Church to ban Copernicanism.

Galileo, worried by this, traveled to Rome to speak to the ecclesiastical authorities. He argued that the Bible was not intended to tell us anything about scientific theories, and that it was usual to assume that, where the Bible conflicted with common sense, it was being allegorical. But the Church was afraid of a scandal that might undermine its fight against Protestantism, and so took repressive measures. It declared Copernicanism "false and erroneous" in 1616, and commanded Galileo never again to "defend or hold" the doctrine. Galileo acquiesced.

In 1623, a longtime friend of Galileo's became the pope. Immediately Galileo tried to get the 1616 decree revoked. He failed, but he did manage to get permission to write a book discussing both Aristotelian and Copernican theories, on two conditions: he would not take sides and would come to the conclusion that man could in any case not determine how the world worked because God could bring about the same effects in ways unimagined by man, who could not place restrictions on God's omnipotence.

The book, *Dialogue Concerning the Two Chief World Systems,* was completed and published in 1632, with the full backing of the censors—and was immediately greeted throughout Europe as a literary and philosophical masterpiece. Soon the pope, realizing that people were seeing the book as a convincing argument in favor of Copernicanism, regretted having allowed its publication. The pope argued that although the book had the official blessing of the censors, Galileo had nevertheless contravened the 1616 decree. He brought Galileo before the Inquisition, who sentenced him to house arrest for life and commanded him to publicly renounce Copernicanism. For a second time, Galileo acquiesced.

Galileo remained a faithful Catholic, but his belief in the independence of science had not been crushed. Four years before his death in 1642, while he was still under house arrest, the manuscript of his second major book was smuggled to a publisher in Holland. It was this work, referred to as *Two New Sciences,* even more than his support for Copernicus, that was to be the genesis of modern physics.

Isaac Newton

Isaac Newton was not a pleasant man. His relations with other academics were notorious, with most of his later life spent embroiled in heated disputes. Following publication of *Principia Mathematica*—surely the most influential book ever written in physics—Newton had risen rapidly into public prominence. He was appointed president of the Royal Society and became the first scientist ever to be knighted.

Newton soon clashed with the Astronomer Royal, John Flamsteed, who had earlier provided Newton with much needed data for *Principia,* but was now withholding information that Newton wanted. Newton would not take no for an answer; he had himself appointed to the governing body of the Royal Observatory and then tried to force immediate publication of the data. Eventually he arranged for Flamsteed's work to be seized and prepared for publication by Flamsteed's mortal enemy, Edmond Halley. But Flamsteed took the case to court and, in the nick of time, won a court order preventing distribution of the stolen work. Newton was incensed and sought his revenge by systematically deleting all references to Flamsteed in later editions of *Principia*.

A more serious dispute arose with the German philosopher Gottfried Leibniz. Both Leibniz and Newton had independently developed a branch of mathematics called calculus, which underlies most of modern physics. Although we now know that Newton discovered calculus years before Leibniz, he published his work much later. A major row ensued over who had been first, with scientists vigorously defending both contenders. It is remarkable, however, that most of the articles appearing in defense of Newton were originally written by his own hand—and only published in the name of friends! As the row grew, Leibniz made the mistake of appealing to the Royal Society to resolve the dispute. Newton, as president, appointed an "impartial" committee to investigate, coincidentally consisting entirely of Newton's friends! But that was not all: Newton then wrote the committee's report himself and had the Royal Society publish it, officially accusing Leibniz of plagiarism. Still unsatisfied, he then wrote an anonymous review of the report in the Royal Society's own periodical. Following the death of Leibniz, Newton is reported to have declared that he had taken great satisfaction in "breaking Leibniz's heart."

During the period of these two disputes, Newton had already left Cambridge and academe. He had been active in anti-Catholic politics at Cambridge, and later in Parliament, and was rewarded eventually with the lucrative post of Warden of the Royal Mint. Here he used his talents for deviousness and vitriol in a more socially acceptable way, successfully conducting a major campaign against counterfeiting, even sending several men to their death on the gallows.

Glossary

absolute zero: The lowest possible temperature, at which a substance contains no heat energy.

acceleration: The rate at which the speed of an object is changing.

anthropic principle: We see the universe the way it is because if it were different, we would not be here to observe it (page 130).

antiparticle: Each type of matter particle has a corresponding antiparticle. When a particle collides with its antiparticle, they annihilate, leaving only energy (page 73).

atom: The basic unit of ordinary matter, made up of a tiny nucleus (consisting of protons and neutrons) surrounded by orbiting electrons (page 63).

big bang: The singularity at the beginning of the universe (page 9–10).

big crunch: The singularity at the end of the universe.

black hole: A region of space-time from which nothing, not even light, can escape, because gravity is so strong (Chapter 6).

Chandrasekhar limit: The maximum possible mass of a stable cold star, above which it must collapse to a black hole (page 88).

conservation of energy: The law of science that states that energy (or its equivalent in mass) can be neither created nor destroyed.

coordinates: Numbers that specify the position of a point in space and time (page 24).

cosmological constant: A mathematical device used by Einstein to give space-time an inbuilt tendency to expand (page 42).

cosmology: The study of the universe as a whole.

electric charge: A property of a particle by which it may repel (or attract) other particles that have a charge of similar (or opposite) sign.

electromagnetic force: The force that arises between particles with electric charge, the second strongest of the four fundamental forces (page 75).

electron: A particle with a negative electric charge that orbits the nucleus of an atom.

electroweak unification energy: The energy (around 100

GeV) above which the distinction between the electromagnetic force and the weak force disappears (page 77).

elementary particle: A particle that, it is believed, cannot be subdivided.

event: A point in space-time, specified by its time and place (page 24).

event horizon: The boundary of a black hole (page 94).

exclusion principle: Two identical spin-1/2 particles cannot have (within the limits set by the uncertainty principle) both the same position and the same velocity (page 72).

field: Something that exists throughout space and time, as opposed to a particle that exists at only one point at a time.

frequency: For a wave, the number of complete cycles per second.

gamma ray: Electromagnetic waves of very short wavelength, produced in radioactive decay or by collisions of elementary particles.

general relativity: Einstein's theory based on the idea that the laws of science should be the same for all observers, no matter how they are moving. It explains the force of gravity in terms of the curvature of a four-dimensional space-time (page 32).

geodesic: The shortest (or longest) path between two points (page 32).

grand unification energy: The energy above which, it is believed, the electromagnetic force, weak force, and strong force become indistinguishable from each other (page 79).

grand unified theory (GUT): A theory that unifies the electro-magnetic, strong, and weak forces (page 79).

imaginary time: Time measured using imaginary numbers (page 141).

light cone: A surface in space-time that marks out the possible directions for light rays passing through a given event (page 28).

light-second (light-year): The distance traveled by light in one second (year).

magnetic field: The field responsible for magnetic forces, now incorporated, along with the electric field, into the electromagnetic field.

mass: The quantity of matter in a body; its inertia, or resistance to acceleration.

microwave background radiation: The radiation from the glowing of the hot early universe, now so greatly red-shifted that it appears not as light, but as microwaves (radio waves with a wavelength of a few centimeters) (pages 43–44).

naked singularity: A space-time singularity not surrounded by a black hole (page 93).

neutrino: An extremely light (possibly massless) elementary matter particle that is affected only by the weak force and gravity.

neutron: An uncharged particle, very similar to the proton, which accounts for roughly half the particles in the nucleus of most atoms (page 68).

neutron star: A cold star, supported by the exclusion principle repulsion between neutrons (page 88).

no boundary condition: The idea that the universe is finite but has no boundary (in imaginary time) (page 142–43).

nuclear fusion: The process in which two nuclei collide and coalesce to form a single, heavier nucleus.

nucleus: The central part of an atom, consisting only of protons and neutrons, held together by the strong force.

particle accelerator: A machine that, using electromagnets, can accelerate moving charged particles, giving them more energy.

phase: For a wave, the position in its cycle at a specified time: a measure of whether it is at a crest, a trough, or at some point between.

photon: A quantum of light.

Planck's quantum principle: The idea that light (or any other classical waves) can be emitted or absorbed only in discrete quanta, whose energy is proportional to their frequency (page 58).

positron: The (positively charged) antiparticle of the electron.

primordial black hole: A black hole created in the very early universe (page 103).

proportional: "X is proportional to Y" means that when Y is multiplied by any number, then so is X. "X is inversely proportional to Y" means that when Y is multiplied by any number, X is divided by that number.

proton: The positively charged particles that make up roughly half the particles in the nucleus of most atoms (page 68).

quantum: The indivisible unit in which waves may be emitted or absorbed (page 58).

quantum mechanics: The theory developed from Planck's quantum principle and Heisenberg's uncertainty principle (Chapter 4).

quark: A (charged) elementary particle that feels the strong force. Protons and neutrons are each composed of three quarks (page 69).

radar: A system using pulsed radio waves to detect the position of objects by measuring the time a single pulse takes to reach the object and be reflected back.

radioactivity: The spontaneous breakdown of one type of atomic nucleus into another.

red shift: The reddening of light from a star that is moving away from us, due to the Doppler effect (pages 40–41).

singularity: A point in space-time at which the space-time curvature becomes infinite (page 50).

singularity theorem: A theorem that shows that a singularity must exist under certain circumstances—in particular, that the universe must have started with a singularity (pages 53–54).

space-time: The four-dimensional space whose points are events (page 25).

spatial dimension: Any of the three dimensions of space-time that are spacelike—that is, any except the time dimension.

special relativity: Einstein's theory based on the idea that the laws of science should be the same for all freely moving observers, no matter what their speed (page 30).

spectrum: The splitting of, say, an electromagnetic wave into its component frequencies (page 40).

spin: An internal property of elementary particles, related to, but not identical to, the everyday concept of spin (pages 70–71).

stationary state: One that is not changing with time: a sphere spinning at a constant rate is stationary because it looks identical at any instant, even though it is not static.

strong force: The strongest of the four fundamental forces, with the shortest range of all. It holds the quarks together within protons and neutrons, and holds the protons and neutrons together to form atoms (page 77).

uncertainty principle: One can never be exactly sure of both the position and the velocity of a particle; the more accurately one knows the one, the less accurately one can know the other (page 58).

virtual particle: In quantum mechanics, a particle that can never be directly detected, but whose existence does have measurable effects (page 73).

wavelength: For a wave, the distance between two adjacent troughs or two adjacent crests.

wave/particle duality: The concept in quantum mechanics that there is no distinction between waves and particles; particles may sometimes behave like waves, and waves like particles (page 61).

weak force: The second weakest of the four fundamental forces, with a very short range. It affects all matter particles, but not force-carrying particles (page 76).

weight: The force exerted on a body by a gravitational field. It is proportional to, but not the same as, its mass.

white dwarf: A stable cold star, supported by the exclusion principle repulsion between electrons (page 88).

Index

A

Abacus, 155
Absolute position, 18
Absolute time, 18, 20, 22, 36, 90, 151
Absolute zero, 193
Acceleration, 193
Age of the universe, 114
Air (element), 10, 67
Air resistance, 16
Albrecht, Andreas, 139
α-particles, 68, 70
Alpher, Ralph, 124
Anthropic principle, 130–133, 140, 145, 153, 159, 174, 193
Antiparticle, 81, 113, 193
Aristotle, 2, 8, 10, 15, 17, 18, 40, 67, 185
On the Heavens, 2
Arrows of time, 151–161, 184
definitions, 160–161
Asymptotic freedom, 78
Atomic structure, 164
Atoms, 63–64, 68, 193
See also Nuclear headings
Augustine. *See* St. Augustine
Axis of symmetry, 96–97

B

Bardeen, Jim, 110
Bekenstein, Jacob, 110, 111
Bell, Jocelyn, 98
Bell Telephone Laboratories, 43
Bentley, Richard, 6
Berkeley, Bishop George, 18
Bethe, Hans, 124
Big bang, 9–10, 50, 54, 66, 105, 121, 131, 140, 183, 194
Big crunch, 121, 183, 194
Black holes, 53, 66, 84, 85–103, 105–120, 121, 158, 183, 194
emissions, 110
primordial, 103, 114–118, 197
properties, 97
rotating, 97, 111
Bohr, Niels, 64
Bondi, Hermann, 50, 51
Born, Max, 164
Brownian motion, 68

C

Calculus, 192
California Institute of Technology, 69, 99, 171

Carbon, 126
Carter, Brandon, 96–97, 110
Case School of Applied Science (Cleveland), 20
Catholic Church, 50, 122
Cerenkov radiation, 118
CERN (European Centre for Nuclear Research), 77
Chadwick, James, 68–69
Chandrasekhar, Subrahmanyan, 87, 88
Chandrasekhar limit, 88, 89, 194
Chaotic boundary conditions, 129
Chaotic inflationary model, 139, 147
City of God, The (St. Augustine), 8
Closed loops, 168
Cold stars, 88
Collapsing universe. *See* Contracting universe
Complete unified theory, 14, 79, 141, 161, 163, 176
  See also Grand Unified theory; Quantum theory of gravity
Confinement, 77
Conservation of energy, 194
Contracting universe, 6, 158, 159
Coordinates, 24, 194
Copernicus, Nicholas, 4, 5, 133
Cosmic censorship, 93
Cosmological arrow of time, 153, 159
Cosmological constant, 135, 139, 159, 194

Cosmology, definition of, 194
Creation, 8, 10, 122, 149, 184
Critical speed, 42
Critique of Pure Reason (Kant), 8
Cronin, J. W., 83
Curvature, space-time, 53, 93, 142, 173
Cygnus X-1, 99

D

Dalton, John, 67
Darwin, Charles, 13
Democritus, 67
Density:
  black holes, 93
  matter particles, 53, 72
  universe, 9, 48–49
Determinism, 57, 59, 182
Dialogue Concerning the Two Chief World Systems (Galileo), 190
Dicke, Bob, 44
Dimensions, 173–175
Dirac, Paul, 59, 72–73, 164
Doppler effect, 40–41, 49
Drexel University, 139

E

Earth (element), 10, 67
Earth (planet), 126
  circumference, 2
  motion, 2
  shape, 2
Eclipses, 2, 19, 34

Eddington, Arthur, 87, 89, 159, 178

Einstein, Albert, 20, 21, 32, 60, 68, 89, 159, 164, 184
  biography, 187–188
  energy equation, 21, 95, 114, 165
  See also General relativity

Elasticity, 164

Electrical force, 19, 175

Electric charge, 194

Electromagnetic field, 19, 70, 112

Electromagnetic force, 75, 163, 194

Electromagnetic waves, 58
  See also Radio waves; Visible light; X rays

Electrons, 63, 64, 68, 75, 112, 118, 123, 164, 177, 194
  volts, 70

Electroweak unification energy, 194–195

Elementary particle, 195

Elements, 10, 40, 67, 126

Energy, 35, 58
  black holes, 115
  conservation of, 194
  particles, 70, 166
  See also Grand unification energy

Enegy equation. *See* Einstein, Albert, energy equation

Entropy, 108, 153
  black holes, 109–110

Ether, 20

Euclid, 142

Euclidean space-time, 142, 143, 145

European Center for Nuclear Research (CERN), 77

Event, 24, 195

Event horizon, 90, 91, 94, 105, 195
  nondecreasing property, 108

Evolution of the universe, 126–127, 130–131

Exclusion principle, 72, 73, 87, 195

Expanding universe, 6, 9, 10, 36, 37–55, 42
  inflationary, 134, 135, 136
  rate, 135

Expansion, critical rate of, 128

F

Feynman, Richard, 65, 120, 141, 142

Field, 195
  See also Electromagnetic field; Magnetic field

Fire (element), 10, 67

"First cause," 8

Fitch, Val, 83

Fixed stars. *See* Stars, fixed

Flamsteed, John, 191, 192

Forces, 74–76

Fourth dimensional space. *See* Space-time

Franklin Institute, 138

Frequency, 195

Friedmann, Alexander, 43, 122

G

Galaxies, 38, 43
Galilei, Galileo, 4, 15, 16, 133
  biography, 189–190
Gamma ray detectors, 118
Gamma rays, 19, 115, 116, 117, 195
Gamow, George, 44, 124
Gell-Mann, Murray, 69
General relativity, 21–22, 33, 34, 35, 36, 43, 48, 50, 53, 65, 84, 86, 87, 90, 94, 121, 128, 135, 140, 143, 156, 164, 184, 195
Genesis, Book of, 8
Geocentric cosmology, 2–3, 133
Geodesic, 32, 33, 34, 195
Geometry, 142
GeV, 76, 79, 177
Glashow, Sheldon, 77
Glueballs, 78
Gluon, 77, 78, 171
Gold, Thomas, 50, 51
Gonville and Caius College (Cambridge), 69
Göttingen University, 164
Grand unification energy, 195
Grand unified theory, 79, 82, 84, 164, 176, 177, 196
Gravitational field, 66, 90, 91, 112, 142
Gravitational force, 74, 84, 99, 171, 175
Gravitational waves, 94
Gravitons, 74–75, 166, 168

Gravity, 5, 6–7, 12, 16–17, 31, 42, 48, 53, 57, 60, 67, 70, 86, 141, 163, 179, 183
  definition of, 10
  quantum effects, 120
  See also Quantum theory of gravity; Supergravity
Green, Mike, 171
GUT. *See* Grand unified theory
Guth, Alan, 134, 136, 137, 139

H

Halley, Edmond, 191
Halliwell, Jonathan, 144
Hartle, Jim, 144
Harvard University, 76
Heat, 126
Heat conduction, 164
Heisenberg, Werner, 58, 59, 60
  See also Uncertainty principle
Heliocentric cosmology, 4, 133
Helium, 125, 126
Herschel, Sir William, 38
Heterotic string, 173
Hewish, Antony, 98
Hoyle, Fred, 50, 51
Hubble, Edwin, 9, 38, 42, 43, 45, 54
Human behavior, 57, 179
Human memory, 155
Hydrogen, 86–87, 125, 126
Hydrogen bomb, 103, 123
Hydrogen sulfide, 127

I

Imaginary time, 142, 147, 151, 196
Imperial College (London), 76
Infinite density, 93
Infinite static universe, 6, 7, 9
Infinity, 6
Inflationary expansion, 134, 135, 136, 148, 157
Initial state of the universe, 12
Interference, 63
Israel, Werner, 95–96

J

Jeans, James, 58
Johnson, Dr. Samuel, 18
Jupiter, 3, 4, 37
    eclipses, 19
    moons of, 19

K

K-mesons, 83
Kant, Immanuel, 8, 185
Kepler, Johannes, 4–5
Kerr, Roy, 96
Khalatnikov, Isaac, 52–53
Kings College (London), 97, 119

L

Laflamme, Raymond, 159
Landau, Lev Davidovich, 88

Laplace, Marquis de, 57, 59, 86, 182
Lebedev Institute, 138
Lee, Tsung-Dao, 82
Leibniz, Gottfried, 192
Levity, 67
Lifshitz, Evgenii, 52–53
Light, 70, 126
    color, 40, 62
    energy, 35
    motion, 25–26
    particle theory of, 85, 86
    propagation of, 19
    speed of, 19–20, 21–22, 25, 30, 85, 86
    visible, 19, 41
    wavelengths, 69, 70
    wave theory of, 85
Light cone, 53, 90, 196
Light-second, 23–24, 196
Light wave, 40, 69, 94
Linde, Andrei, 138, 139, 147
Lord Rayleigh, 58
Lorentz, Hendrick, 20
Lucasian Professorship of Mathematics (Cambridge), 72
Luttrel, Julian, 144

M

Macromolecules, 127
Magellanic clouds, 100
Magnetic field, 100, 196
Magnetism, 19
Mars, 3, 37
Mass, 16–17, 196
    black holes, 97
    See also Chandrasekhar limit

Massachusetts Institute of Technology, 134
Matter, 67
  density, 53
  properties of, 164
Matter particles, 72, 88, 112, 136
Maxwell, James Clerk, 19, 76
Measurement, 22, 58–59
Mercury (planet), 3, 11, 34
Mesons, 77
  See also K-mesons
Michell, John, 85–86, 99
Michelson, Albert, 20
Microelectronics, 14
Microwave background radiation, 44, 51, 114, 148, 196
Microwaves, 19
  temperature, 127
Milky Way, 38, 43
Molecules, 65, 67–68
Moons, eclipses, 2, 19
Morley, Edward, 20
Moss, Ian, 138
Motion, 15, 35, 36
  Brownian, 68
  See also Orbits
Mott, Nevill, 69

N

Naked singularity, 196
Natural selection, 13
Nature of the universe, 10
Neutrinos, 123, 196
Neutrons, 68, 123, 164, 169, 196

Neutron stars, 88, 95, 99, 196
Newton, Isaac, 5, 6, 10, 16, 18, 35, 85
  biography, 191–192
  Philosophiae Naturalis Principia Mathematica, 5
  See also Gravity
Nobel prize:
  Born, 164
  Chadwick, 69
  Chandrasekhar, 89
  Dirac, 73
  Einstein, 60
  Fitch, 83
  Gell-Mann, 69
  Glashow, 77
  Lee, 82
  Michelson, 20
  Mott, 69
  Pauli, 72
  Penzias, 45
  Rubbia, 77
  Salam, 77
  Van der Meer, 77
  Weinberg, 77
  Wilson, 45
  Yang, 82
No boundary condition, 144, 148, 153, 157, 197
No hair theorem, 97
North Star, 2
Nuclear energy, 14
Nuclear force, 76, 77, 163, 164
Nuclear fusion, 126, 197
Nucleus, 197
Numbers:
  specifying time, 168
  value, 132

O

Occam's razor, 59
Olbers, Heinrich, 7
On the Heavens (Aristotle), 2
Open strings, 168
Oppenheimer, Robert, 89
Orbits, 2, 4, 5, 17, 175
Origin of the universe, 7–10,
    36, 121–149
    See also Creation
Oxygen, 126

P

Page, Don, 159
Palomar Observatory, Califor-
    nia, 97
Particle accelerator, 197
Particles, 67–84, 88, 141, 177
    light, 85
    position, 58, 112, 183
    velocity, 112, 183
    See also Elementary
        particle; Matter parti-
        cles; Virtual Particles;
        Waves
Pauli, Wolfgang, 72
    See also Exclusion principle
Peebles, Jim, 44
Penn State University, 159
Penrose, Roger, 36, 53, 93,
    96, 140
Penzias, Arno, 43–44, 51,
    124
Phase, 197
Phase transition, 134
*Philosophiae Naturalis Prin-
    cipia Mathematica. See
    Principia Mathematica*

*Philosophical Transactions of
    the Royal Society of
    London,* 86
Philosophers, 185
Photons, 74, 75–76, 86, 123,
    197
Physics, unification of,
    163–179
Planck, Max, 58
Planck's constant, 59
Planck's quantum principle,
    117, 177, 197
Planets, 3, 37, 126
    See also specific names
Pluto, 117
Poincaré, Henri, 20
Popper, Karl, 11
Porter, Neil, 118
Positrons, 73, 118, 197
Primordial black holes, 103,
    114–118, 197
Princeton University, 44
Principia Mathematicia (New-
    ton), 5, 16, 19, 191
Prism, 40
Proportional, 197
Protons, 68, 123, 164, 169,
    197
    spontaneous decay, 80–81
Proxima Centauri, 37
Psychological arrow of time,
    153, 155–156
Ptolemy, 4, 40, 133
Pulsars, 95, 98

Q

Quantum, 58, 198
Quantum gravitational effects,
    120, 141

Quantum mechanics, 48, 54, 55, 59–61, 65, 70, 71, 84, 121, 141, 164, 165, 176, 182, 198
definition of, 13
Quantum theory of gravity, 13, 84, 137, 141, 143, 156
See also Complete unified theory
Quarks, 69, 75, 77–78, 81, 112, 171, 177, 198
Quasars, 98, 100
Queen Mary College (London), 171

R

Radar, 50, 198
Radiation, 58, 81, 124
black holes, 110, 111, 119
See also Microwaves
Radioactivity, 76, 198
Radio waves, 19, 51, 95, 98, 100
Rate of expansion, 128
Real time, 152
Reason, pure, 8
Reasoning abilities, 14
Red shift, 41, 198
Relative time, 151
Relativity. *See* General relativity; Special relativity
Renormalization, 165
Rest, state of, 17
Robertson, Howard, 45

Robinson, David, 97
Roemer, Ole Christensen, 19, 85
Roman Catholic Church. *See* Catholic Church
Rubbia, Carlo, 77
Russell, Bertrand, 1
Rutherford, Ernest, 68, 70
Ryle, Martin, 51

S

St. Augustine, 8, 9, 176
Salam, Abdus, 76, 77
Saturn, 3, 37
Scherk, Joël, 171
Schmidt, Maarten, 97
Schrödinger, Erwin, 59
Schwarz, John, 171
Schwarzchild, Karl, 95
Science fiction, 173
Scientific goals, 11
Scientific laws, 129
Scientific theory, 11
definition of, 10, 147
Singularity, 53, 93, 120, 140, 141, 147, 165, 183, 198
See also Naked singularity
Singularity theorem, 50, 54, 66, 140, 198
Sirius, 88
Slow symmetry breaking, 138
Solar system, state of, 57
Sound waves, 20
Space, 15–36
absolute, 18
properties of, 48

Space-time, 25, 121, 142, 198
  curvature, 53, 93, 142, 173
  dimensions, 173–176
  life in, 175
  line in, 168
  See also Euclidean space-time
Space travel, 173
Spatial dimension, 198
Spatial movement, 5
Special relativity, 31, 36, 72, 198
Spectrum, 40, 199
Speed of light, 19–20, 21–22, 25, 30, 85, 86
Spin, 70–71, 166, 199
Spiral galaxy, 38
Spontaneous proton decay, 80–81
Spontaneous symmetry breaking, 76
Starobinsky, Alexander, 111
Stars:
  cataloging of, 38
  collapsing, 53, 90–91
  fixed, 3, 5
  life cycles of, 86
  luminosity of, 38
  number of, 6
  temperature of, 40, 124
  visible, 37
  See also Cold Stars; Neutron stars; North Star
Stationary state, 199
Steady state theory, 50, 51
Steinhardt, Paul, 139
Sternberg Astronomical Institute, 137
String theories, 168–173

Strong anthropic principle, 131–133
Strong force, 169, 171, 199
Strong nuclear force, 77
Sun:
  eclipse, 34
  motion, 4
  temperature, 123
Supergravity, 166, 167–168, 171
Supernova, 126
Symmetry, 76, 82, 137, 152
  axis of, 96
  See also Slow symmetry breaking
System of the World, The (Laplace), 86

T

Taylor, John G., 119
Telescopes, 4
Temperature:
  microwaves, 127
  stars, 124
  sun, 123
Thermal spectra, 40
Thermodynamic arrow of time, 153, 154, 155, 156, 160–161
Thermodynamics, second law of, 108–109, 110, 111, 153
Third dimension. *See* Dimensions
Thomson, J.J., 68, 70
Thorne, Kip, 99
Time, 8–9, 15–36, 151–161
  absolute, 18, 20, 22, 36, 90, 151

Time (*continued*)
 arrows of, 153
 imaginary, 141, 147, 151,
   152, 196
 properties of, 48
Trinity College, Cambridge,
   68
Two New Sciences (Galileo),
   190

U

Ultraviolet light, 19
Uncertainty principle, 48,
   57–66, 72, 111, 148,
   157, 164, 165, 182,
   199
Unchanging universe, 10
Unification of physics. *See*
   Physics, unification of
Universal gravitation. *See*
   Gravity
Universal interdependence,
   12
Universe:
 age of, 114
 contracting, 6, 158, 159
 density, 9, 48–49
 evolution of, 126–127,
   130–131
 infinite static, 6, 7,
   9
 initial state, 12
 See also Expanding
   universe; Origin of the
   universe
University of California
   (Santa Barbara), 144
University of Pennsylvania,
   139

V

Van der Meer, Simon, 77
Vector bosons, 76
Velocities, 52
Velocity, 58, 59, 88, 183
 of particles, 73, 112,
   183
Venus, 3, 37
Virtual particles, 112, 166,
   199
Virtual photons. *See*
   Photons
Visible light, 19, 41

W

Walker, Arthur, 45
Water, 10, 67
Wavelengths, 19, 199
 of light, 69, 70
Wave/particle duality, 65, 85,
   199
Waves, 74
 gravitational, 94
 light, 85
 See also Particles
Weak anthropic principle,
   130–131, 140, 153,
   159
Weak force, 199
Weak nuclear force, 76
Weekes, Trevor, 118
Weight, 200
Weinberg, Steven, 76
Wheeler, John, 85, 96, 101
White dwarf, 88, 99, 200
Wilson, Robert, 43–44, 51,
   124
Wittgenstein, Ludwig, 185

"World-line," 168
World-sheet, 168
Wu, Chien-Shiung, 82

X

X rays, 19, 115

Y

Yang, Chen Ning, 82

Z

Zeldovich, Yakov, 111

About the Author

STEPHEN W. HAWKING is forty-six years old. He was born on the anniversary of Galileo's death, holds Newton's chair as Lucasian Professor of Mathematics at Cambridge University and is widely regarded as the most brilliant theoretical physicist since Einstein.